M

Achieving World Class Manufacturing Through Process Control

Achieving World Class Manufacturing Through Process Control

Joseph P. Shunta

Prentice Hall PTR
Englewood Cliffs, New Jersey 07632

Library of Congress Cataloging-in-Publication

Shunta, Joseph P.
 Achieving world class manufacturing through process control /
Joseph P. Shunta.
  p. cm.
 Includes bibliographical references and index.
 ISBN 0-13-309030-2
 1. Process control. I. Title.
TS156.8.S54 1995
670.42—dc20

94-14213
CIP

© 1995 Prentice Hall PTR
Prentice-Hall, Inc.
A Paramount Communications Co.
Englewood Cliffs, New Jersey 07632

Editorial / production supervision: *Mary P. Rottino*
Cover design: *Karen Marsilio*
Acquisitions editors: *Michael Hays / Bernard Goodwin*

The publisher offers discounts on this book when ordered in bulk quantities.
For more information contact:

Corporate Sales Department
P T R Prentice Hall
113 Sylvan Avenue
Englewood Cliffs, NJ 07632.

Phone: 201-592-2863
FAX: 201-592-2249

Printed in the United States of America

10 9 8 7 6 5 4 3 2 1

ISBN 0-13-309030-2

Prentice-Hall International (UK) Limited, *London*
Prentice-Hall of Australia Pty. Limited, *Sydney*
Prentice-Hall Canada Inc., *Toronto*
Prentice-Hall Hispanoamericana, S.A., *Mexico*
Prentice-Hall of India Private Limited, *New Delhi*
Prentice-Hall of Japan, Inc., *Tokyo*
Simon & Schuster Asia Pte. Ltd., *Singapore*
Editora Prentice-Hall do Brasil, Ltda., *Rio de Janeiro*

*To my parents for their love,
encouragement, and support
over these many years*

Contents

*statistical
and tools to
quantify variability
and how to
improve it!*

Contents

How to analyzes the process and apply the tool to identify process control opportunities and estimate the benefits.

Preface

More than ever before, global competition is driving companies to make major adjustments in manufacturing to reduce costs and raise quality. Every aspect is being scrutinized, from the point of receiving raw materials to shipping the final product. Companies are making greater use of metrics to monitor and drive improvements in key operating parameters like quality, throughput, first-pass first-quality yield, cycle time, and uptime.

Process control is one of a number of key manufacturing technologies being applied to obtain improvements in operating performance. Many companies recognize that they have a huge economic stake in doing so. The improvements that may provide benefits cover the gamut from simply correcting controller tuning to installing advanced model-based controls. It often happens that there are more worthwhile improvements than there are people to implement them. We have to work "smarter": seeking the opportunities that will have the biggest impact on the business as well as be feasible in terms of cost and timing.

The subject of this book is the improvement of manufacturing through process control. It is intended for managers, engineers, applied statisticians, and quality specialists who are looking for ways to improve their manufacturing processes. Hopefully, they will find helpful guidance on practical ways to accomplish their goals. It presents a methodology that uncovers the significant opportunities for improving process control which are focused on meeting business objectives. The book presents some fresh approaches and tools that will hopefully make this endeavor straightforward and effective. This is followed by showing some practical ways in which automatic controls can be applied to reduce variability. It is not the intent of this book to provide the theory for the statistics or process control strategies presented, so references are given throughout for those wishing to get into the subject more deeply. Although written from the perspective of the chemical process industry, the concepts apply to many types of manufacturing processes, such as food, drug, paper, rubber, plastics, and petrochemicals.

The first chapter discusses how process control, through controlling variability, impacts key business drivers in manufacturing. Chapters 2 and 3 present statistical metrics and tools to quantify variability, and suggest how to improve it. Chapters 4 and 5 show how to analyze the process and apply the tools to identify process control opportunities and estimate the benefits for implementing them. Chapter 6 discusses an approach for prioritizing opportunities based on scarce process control resources. Chapter 7 discusses how automatic process controls are applied to reduce variability and some of the limitations. Chapter 8 is an overview of some mathematical techniques used to predict variables that cannot be measured frequently enough, or maybe at all. Chapter 9 discusses the very important subject of how to sustain the benefits once they are achieved. Finally, Chapter 10 discusses how process design can be influenced to enhance controllability.

The book is based on work my colleagues and I have been engaged in over the past ten or so years to help improve chemical manufacturing through process control. I wrote the book because I believe strongly in the benefits of process control and want to encourage others to discover them. I particularly hope that manufacturing and plant technical resources will find the book encouraging and helpful, because this is where the effort to improve control must be initiated and sustained.

I have worked with and learned from too many folks to list here, but I want to acknowledge some key people in DuPont who have had the most impact on the subject matter of this book. Bill Fellner, Steve Bailey, Pat DeFeo, and Perce Ness educated me on many of the statistical tools used throughout the book. Pat DeFeo, Bill Fellner, and Dave Schnelle were particularly generous in reviewing and making suggestions on the writing. I also want to acknowledge that Bob Cox and John Anderson originated the concept of the Project Selection Index, which is the subject of Chapter 6. Finally, special thanks to Page S. Buckley, who encouraged me to write, and to my wife Beverly, for her support and patience during the many hours I spent at the computer.

Joseph P. Shunta
Newark, Delaware

1 — methodology and approach

2 —

About the Author

Joseph P. Shunta is a Principal Consultant in DuPont Engineering where he has responsibility for dynamic analysis, simulation, and designing process control strategies for new and modernized plants. He joined DuPont in 1964 after receiving a B.S. degree in chemical engineering from Michigan Technological University. After working several years in various technical and manufacturing assignments, he entered Lehigh University and earned an M.S. and Ph.D. in chemical engineering. He rejoined DuPont, and has specialized in process control for the past 20 years. He has written over 25 papers and co-authored the book *Design of Distillation Column Control Systems* with Page S. Buckley and William L. Luyben. He is a registered professional engineer in Delaware and California.

Introduction

BECOMING "WORLD CLASS" IN MANUFACTURING

We hear many things referred to these days as "world class." Whether the term is used to describe cars, athletes, or something else, it means being among the best in the world. The performance achieved by world-class products and people are the standards for global competition. Manufacturing companies have also adopted the term to describe processes, products, and services that achieve standards of performance among the best in the world. Companies that are global players must strive to reach and sustain those standards if they are to survive the fierce competitive pressures of the world markets.

Manufacturing companies are judged by the quality, cost, and availability of the products they sell. To achieve world class performance in these areas, the manufacturing operation must develop specific manufacturing metrics to assess how well they are performing day to day and guide them towards improvements. Among the specific metrics being used today are process capability indices (Cp/Cpk) and process performance indices (Pp/Ppk) to describe quality, % first-pass first-quality yield, % uptime, and cycle time. Performing to world class standards in these specific metrics leads to competitive quality, cost, and availability of their products.

The task of achieving world-class performance is illustrated in Figure 1–1. It is to drive performance, using people and technology, to world-class standards for that particular business. It takes both people and technology to reach the goal. Leadership defines and communicates the goals and tasks, and provides a manufacturing system and environment in which the goals can be achieved. Employees must be empowered by the leadership to develop and apply their skills to effectively operate, maintain, and support the manufacturing process.

Technology is the means to transform raw materials into saleable products. It includes the process equipment, measurements, and controls. Technology alone cannot ensure world class performance, no matter how sophisticated. This point was vividly driven home by the following actual situation. Company A was being beaten out in the marketplace by Company B, which produced the same product. Company B had comparable equipment but not much automation. They did have a competent and motivated workforce. Company A decided the way to beat Company B was to install a computer control system that would consistently

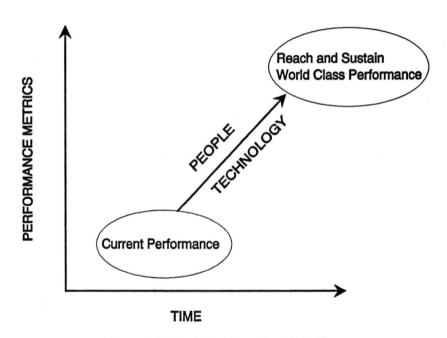

Figure 1-1   The Path to Becoming World-Class

make the product to exacting tolerances at a higher throughput rate than Company B. However, even after the new system was operational, Company B still maintained its lead. Company A started an investigation and soon discovered the problem. Whenever the computer detected that a limit had been violated or some malfunction had occurred, it shut down the line and signaled the operator. Seldom was someone around to fix the problem and get the unit on line again, so valuable production time was lost.

The moral of this story is that smart technology alone cannot ensure success without proper operating procedures and trained, motivated personnel. Even if the latest and greatest measurement and control systems are installed, the investment will not return its full potential unless control strategies are designed to achieve the desired performance and maintained to sustain the benefits.

PROCESS CONTROL AND BUSINESS GOALS

Process control is a key technology to help drive manufacturing plants to world class performance. Once the process equipment is installed and running, it follows that how it is controlled determines how well it performs.

What do we mean by process control? Process control means different things to different people. To some, it may mean control equipment like distributed control systems, smart transmitters, and high-performance valves. To others, it may mean optimum control strategies, statistical control charts, and so on. In reality, it is all these things and more.

Process control is the effective combination of people and technology for maintaining the process at the desired steady-state operating level and maintaining acceptable variability of the process variables and product properties.

By process variables, we mean the physical state of the process streams or system. Examples are flow rate, pressure, temperature, level, density, and concentration. Product properties are the physical attributes of the product that relate to customer requirements for which a defined test method exists. Examples are molecular weight distribution, color, denier, elasticity, viscosity, and purity.

The focal point for improving manufacturing performance should be the business needs or goals. We often refer to these as the *business drivers* because they are the things that drive improvements. Since the job of process control is to manage variability, we must understand the connection between process variability and business goals before we can appreciate the value of process control. Some important business drivers are described below.

Quality, often characterized in terms of process capability (Cp/Cpk) or process performance (Pp/Ppk) indices, is concerned with making a product that fits the customer's needs. Customers want the product properties to conform to their specifications and/or have a specified consistency [Crosby 1979]. Thus, controlling quality is strongly related to controlling product variability. (These metrics are explained in detail in Chapter 3.)

First-pass first-quality yield refers to making products right the first time through the process. If the product conforms to the specifications the first time, rework, blending, waste, or selling product at reduced prices are avoided. Again, first-pass first-quality yield depends to a large extent on controlling variability in the process.

Throughput is the rate at which product is made or materials flow through the process. It is also impacted by variability. Lower variability means that the process can be operated closer to physical constraints to maximize throughput. Indirectly, throughput is impacted by higher first-pass yield because less off-spec material has to be reworked or recycled, either of which may tie up equipment that otherwise could be making fresh product.

Uptime, the amount of time the equipment is available to run at full rate, is enhanced by not having to shut down because of fouling, plugging, exceeding safety interlocks, and so on. Controlling variability makes it easier to avoid troublesome process conditions that cause these problems.

Finally, cycle time, the elapsed time between receiving raw materials and shipping the corresponding product, is shortened by smaller variability. Transitioning between product grades can be accomplished faster, and fewer off-grade products are produced, thus reducing the time it takes to rework them.

While describing these metrics, the focus has been on the impact of controlling variability, which is the primary purpose of a control system. For quality variables (i.e., product properties), the variability has to conform to the customer specifications. The product property variability specifications may be relatively tight. On the other hand, the variability of a process variable depends on what that variable is. For example, the ingredients fed to a reactor often have to be tightly controlled to minimize producing by-products. However, it is usually not critical to minimize the variability of a tank liquid level. In fact, it may be preferable to let the level vary over some range to provide flow-smoothing out of the tank so that downstream equipment is not upset. This may be necessary to produce quality product. Thus, for one variable it is important to minimize variability, while for another it is better to increase variability. *The controls actually shift variability away from where it matters to where it does not matter.*

The stake in controlling variability correctly is that product is made right the first time. This avoids:

- Storing off-quality product, which increases inventories and cycle time
- Blending various off-spec products to meet customer specifications
- Reworking product, which reduces plant capacity, throughput, and energy efficiency
- Selling at a reduced price
- Disposing of off-grade product and other wastes

Reducing variability will also enable process economic optimization by operating closer to constraints [Latour 1986]. The constraint might be the operating point that results in the lowest energy consumption or lowest product giveaway, but beyond which the process should not be operated for other reasons. Figure 1–2 shows a variable with an initially large variability. The operator conservatively sets the target far enough away from the constraint so that it is not ex-

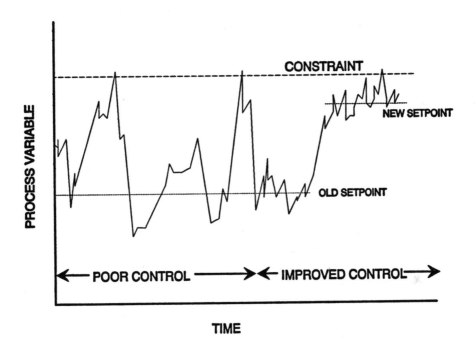

Figure 1–2 Process Optimization by Reducing Variability

ceeded more than a certain amount. By improving control, the variability is reduced, and the operator can move the target closer to the constraint and still not
incur any more violations of the constraint than before.

Figure 1–3 shows two benefits for reducing variability in terms of histograms that plot the value of a product property or process variable versus the
frequency of occurrence. The upper histogram illustrates again how it is possible
to move the average value closer to a limit or constraint (X_L) when the variability
is reduced, that is, the spread in the data is narrowed. The bottom histogram shows
another benefit, namely, that when the variability is reduced, the specifications are
exceeded less frequently. Some examples will be shown later in the book.

SOURCES OF VARIABILITY

The types of upsets and the corresponding variability in the process should be understood in order to effectively apply process control. Juran (1980) describes two
causes of variability in the process.

Common causes are frequent, short-term, random disturbances that are inherent in every process. Examples are small, random variations in steam supply
pressure, vibrations and turbulence, and slight randomness in raw material composition. These kinds of upsets are always present to some extent and cause some
variability, even with excellent control. It may be possible, however, to eliminate
some of these causes if the process or control equipment is changed. For example, we might be able to install a measurement device with greater sensitivity and
speed of response. Often, we are willing to live with these kinds of disturbances
as long as we can still make satisfactory product. Trying to eliminate them might
also be prohibitively expensive relative to the improvement.

Special or assignable causes are larger, less frequent disturbances that are
identifiable and hopefully preventable. Examples are things like equipment malfunctions, operator blunders, gross upsets in steam pressure, fouling, catalyst
decay, and grossly defective batches of raw materials. Process control is used to
compensate for these kinds of upsets.

There are three means available to reduce variability:

- Make frequent compensatory adjustments to the process based on a deviation from the desired value.
- Monitor the key variables to signal the occurrence of a special cause; then
  identify and eliminate the cause.
- Change the process.

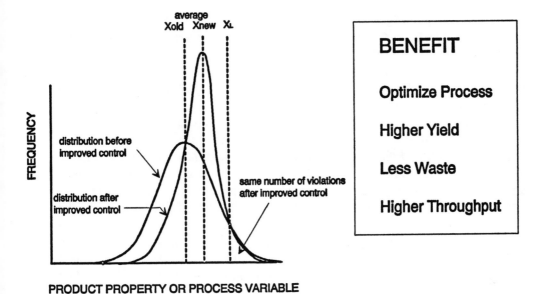

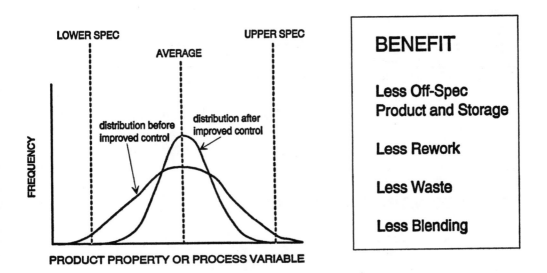

Figure 1–3   Benefits of Reduced Variability

Automatic process control (APC), or regulatory control, tries to compensate for causes or upsets by making frequent adjustments to the process that counteract their effects. APC is particularly aimed at special-cause variability, but may also be effective for long-term random variations as well. Its effectiveness is dependent on the speed of response of the controls, the suitability of the measurement, the control strategy, and so on.

Statistical process control (SPC) seeks to identify and remove root causes for the special-cause variability. Commonly, operators use control charts (e.g., Shewart and CUSUM charts) to do this. The data are plotted on the charts along with calculated control limits. If the data lie between the control limits, the variability is random, and the process is said to be in a state of *statistical process control*. If the data wander outside the control limits, a set of rules determines the point at which the process is out of statistical control and a special cause has occurred [Yeager and Davis 1992]. Although the classic role of SPC is to seek out and eliminate root causes, often SPC is implemented in terms of making a process adjustment to compensate for the special cause and bring the product back within limits. The adjustment rules are usually arbitrary, and the adjustments are done manually.

A combination of APC and SPC is the best way to control variability. If APC and SPC are effective, the resulting variability will mostly be the short-term, random kind inherent in the process. This is about the minimum variability that can be achieved with a given process. If the magnitude of the random variability is too great, the process, or perhaps the control equipment, will have to be changed.

There has been a lot of research activity lately on how to combine APC and SPC effectively [Box and Kramer 1992, Vander Wiel et. al. 1992]. Basically, the trend is to apply APC to compensate for upsets automatically and SPC to:

- Monitor the process variables to detect when APC is not performing effectively
- Monitor the APC output to detect the presence of a special cause

VARIOUS WAYS TO REDUCE VARIABILITY

Some specific ways to reduce variability are listed below:

- Tighten controller tuning on product quality loops
- Eliminate measurement delays and process deadtimes

- Increase sampling frequency
- Decouple interacting control loops
- Apply feedforward, cascade, and multivariable control strategies
- Install on-line measurements
- Use on-line predictive models (inferential sensors)
- Automate startups
- Attenuate or eliminate disturbances
- Simplify the process
- Eliminate the root causes of variability.

Examples of how these reduce variability will be illustrated later in the book.

ACHIEVING BENEFITS THROUGH PROCESS CONTROL

Having become convinced that process control is a technology that can drive the process to world class performance, we have to develop a strategy for accomplishing it. Such a strategy is illustrated in Figure 1–4.

First, it is necessary to identify the opportunities that will have a significant

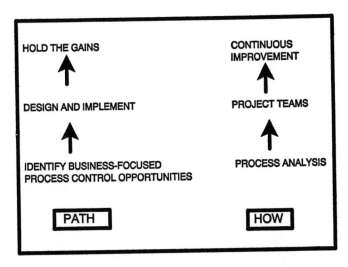

Figure 1–4   Achieving Benefits Through Process Control

impact on the business. This book presents an analysis methodology for doing this.

Second, the opportunities are converted into actual hardware and software improvements and then implemented. This task is executed by a team with the required skills and project experience. The scope of the project is defined by the results of the first step.

Last, it is necessary to sustain or hold the gains achieved. Holding the gains comes by making continuous improvements to the process and systems for monitoring and controlling the process. This requires that a plant infrastructure be in place with the skills and empowerment to carry this out. This issue will be addressed later.

The first part of the book addresses the first task in the strategy: identifying the business-focused process control opportunities. A process analysis methodology is presented that can be applied to:

- Assess current operation in terms of the performance metrics
- Identify the process measurement and control opportunities
- Estimate how much performance can be improved through control

Once this is done, the opportunities can be prioritized based on their relative impact on the business and the feasibility of implementing them. Some of the opportunities will be like "low-hanging fruit," easy to get. These can be attacked first with relatively little effort while getting a quick return. Others will require more effort, but will still be worthwhile because of benefits to the business. Long-term opportunities will be those having high impact but low feasibility, perhaps requiring research and development. Whatever the case may be, the analysis provides a straightforward and logical strategy to identify opportunities for improving manufacturing.

Figure 1–5 illustrates the overall steps in the process analysis phase. Identifying opportunities begins with stating the important business goals or drivers. This provides the focus needed to address the right issues. Unless the business drivers are addressed, the results may be tough to sell to the business; also, the improvements may not achieve the desired results, even though they might be interesting or challenging. Clearly, productivity will be improved when the investment in capital and resources is for things that will affect the bottom line.

Next, the product properties and process variables that impact the business drivers are identified. Obviously, these are the ones we should be monitoring and controlling well. Of all the variables in a process, these key ones most need to be

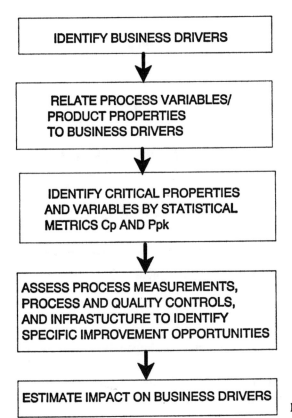

Figure 1-5   Process Analysis Steps

investigated. Some others may be troublesome or a nuisance; but unless they have a direct impact on quality or first-pass yield or some other business driver of concern, they really should not be given top priority.

Statistical metrics (process capability Cp and process performance Ppk), are applied to determine which of the key properties and variables do not meet the desired performance. The statistics provide a straightforward way to determine if something in the control strategy needs to be modified or changes in the process need to be made to gain the improvements. This step identifies the gems for improvement.

The systems for measuring and controlling the properties and variables and their support systems are assessed to identify barriers to good performance and opportunities for making improvements. In particular, the systems applied to the variables and product properties that are not performing in a satisfactory manner are scrutinized. These opportunities are candidates for implementation after they

---

Identifies Business-Focused Improvements

Provides Focus and Direction for Implementation

Provides Basis for Expectations and Achieving Results

Provides Basis for Continuous Improvement

Figure 1-6   Benefits of Process Analysis

have been found to be feasible and have enough of an impact on the business that it makes sense to implement them.

Finally, the impact that the improvements will have on the business is estimated. These data are used to justify the implementation of the improvements.

The benefits of all this effort are listed in Figure 1–6. First of all, it draws attention to business-focused improvements. Improvement programs often lack business focus. As a result, they may use skilled resources to solve problems that will have no significant effect on the bottom line. The process analysis provides a framework and direction for implementing the changes that really count. Estimating a stake creates a tangible goal and expectations that everyone can relate to and hopefully buy into. Finally, the analysis can be repeated as often as necessary to continue the improvement cycle so that the gains are sustained.

SELECTING THE PROCESS

Before going into an analysis, it would be nice to have some confidence beforehand that process control benefits may be significant. This section presents a screening technique to rapidly check your process characteristics against those that would make the process qualify for analysis. You simply go through the checklist and pick the items that apply to your process. If you check off a number of items, odds are good that an analysis will be worthwhile.

In general, the definition below describes the types of processes where automatic process control systems are required to gain benefits:

> Continuous or batch wet chemical processes where the dynamic response of variables is important. These processes typically use PID (proportional-integral-derivative) controllers. Operations not fitting this description include strictly mechanical operations where dynamics are not an issue. These are typically controlled by discrete logic controllers, for example, programmable logic controllers.

Although this definition directly addresses the chemical process industry, it is appropriate for many other kinds of processes in which process control is applied to achieve dynamic compensation.

BUSINESS OBJECTIVES

Process control plays a major part in achieving many business objectives. The following lists the kinds of concerns around various business objectives that usually call for high process control performance. If any number of these apply and are significant enough to affect profitability, there may be opportunities for improved process control.

QUALITY

_____	Very tight specifications on intermediate or final products
_____	Widely varying feedstock quality
_____	Multiple product grades
_____	Product degradation
_____	Precise recipes
_____	Low C_p, P_p
_____	Large product price differentials

THROUGHPUT

_____	Excessive turndown requirements (5:1 or higher)
_____	Frequent startups and shutdowns
_____	Process bottlenecks
_____	Frequent product transitions

_____   High uptime requirements (90% or higher)
_____   High percentage of rework
_____   Operation near equipment constraints (95% capacity or
             higher)

YIELD

_____   Low first-pass first-quality yield
_____   Excessive waste
_____   Frequent or difficult product transitions
_____   Costly raw materials

ENVIRONMENTAL

_____   Highly toxic or corrosive chemicals
_____   Large waste streams requiring EPA permits
_____   OSHA-regulated chemicals
_____   Neutralization, reaction, or dilution of waste streams
_____   Variable rate scrubber systems for vent streams

ENERGY

_____   Significant heat integration strategies
_____   Excessive cost of utilities
_____   Complex energy distribution systems

UPTIME

_____   Frequent violation of interlocks or process constraints
_____   Frequent prolonged, difficult startups
_____   Excessive flooding, corrosion, vibration, turbulence in
             process equipment

PROCESS OPERATIONS

This section describes the kinds of chemical unit operations or equipment requiring automatic process control to maintain specified operating levels and control variability. These have a high potential benefit for improved process control.

Reaction Systems: Chemical Reactors, Neutralizers, and Waste Treatment

Much product variability is associated with reaction processes. They are typically nonlinear and require tight control of variability. Reactions may also pose safety concerns around toxicity, flammability, and so on.

Separation Systems: Distillation, Absorption, Extraction, Decantation, Scrubbing, Filtration, Centrifugation, and Evaporation

These systems are often multivariable, that is, they have several control loops that may interact, causing cycling. These systems are often connected in series (cascaded), which may amplify the effect of upsets if not controlled properly.

Energy Exchange Systems: Heaters, Boilers, Furnaces, Dryers, Heat Exchangers, and Vaporizers

Process variability and physical constraints are often of concern in these systems. They are also multivariable, which causes control interactions and cycling, especially heat integration systems.

Flow Systems: Compressors, Pumping, Piping, and Recycles

These devices are critical to keeping the process running. They may be sources of process upsets if not controlled properly. Compressors often require sophisticated anti-surge controls. Complex piping and recycle systems that tie parts of the process together may be the source of instability.

REFERENCES

Box, G. E. P. and Kramer, T. Statistical Monitoring and Feedback Adjustment—A Discussion. *Technometrics*, Vol. 34, No. 3, pp. 251–306, August 1992.

Bozenhardt, H. and Dybeck, M. Estimating Savings from Upgrading Process Control. *Chemical Engineering*, pp. 99–102, February 3, 1986.

Crosby, P. B. *Quality Is Free*. New York: McGraw-Hill, 1979.

JURAN, J. M. and GRYNA, F. M., Jr. *Quality Planning and Analysis.* New York: McGraw-Hill, 1980.

LATOUR, P. L., et. al. Estimating Benefits from Advanced Control. *ISATransactions,* Vol. 25, No. 4, pp. 13–21, 1986.

VANDER WIEL, S. A., et. al. Algorithmic Statistical Process Control: Concepts and an Application. *Technometrics,* Vol. 34, No. 3, pp. 286–297, 1992.

WADSWORTH, H. M. Jr, et. al. *Modern Methods for Quality Control and Improvement.* New York: John Wiley, 1986.

YEAGER, R. L. and DAVIS, T. R. Reduce Process Variation with Real-Time SPC. *Hydrocarbon Processing,* pp. 89–92, March 1992.

2

Assessing Variability

Chapter 1 made the claim that the primary concern of process control is controlling variability in product properties and process variables. Since this book deals with estimating variability and how much it can be reduced, this chapter lays some groundwork by presenting common ways to assess variability. For this, the world of the statisticians is entered. Most of what is covered is very fundamental to a statistician, but may be new to most chemical engineers or control engineers. Later, these statistical metrics are applied to determine the current performance of the controls, estimate the potential reduction in variability, and calculate the benefits from improved control.

ESTIMATING VARIABILITY

It would be nice to have a way to visualize variability that gives a clear picture of its magnitude and form. We can certainly observe variability from trend charts to get a rough appreciation of the frequency and magnitude of the variability rela-

tive to the average. Trend charts also show how the process average shifts or cycles, when some large upsets occur, and when the process is in or out of control.
Thus, trend charts are very helpful in interpreting what is going on in the process.
However, it is difficult to quantify variability from trend charts. For this, there are
a number of specific statistical tools. One of these tools is the frequency distribution.

Figure 2–1 shows a frequency distribution. Data values are plotted on the
x-axis, and the frequency or number of times that value appears in the data set on
the *y*-axis. The distribution shows the variability or spread in the data about the
average. The shape of the distribution is important. Figure 2–1 illustrates a normal or Gaussian distribution, which results when the data are independent (not
correlated) and have the same mean and standard deviation. Such a process is in
statistical process control, that is, it is stable and predictable because it is undis-
turbed by extraneous or special causes. A normal distribution has the property
that the mean or average of the data equals the median or middlemost value.

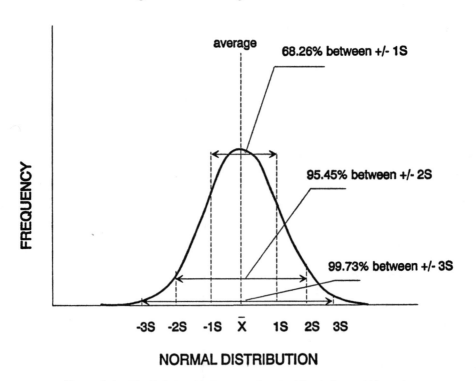

Figure 2–1   The Relationship Between Standard Deviation and Normal
Distribution

Thus, the curve is symmetrical about the average and has the well-known bell curve shape.

The spread of the data is the amount of variation about the average. The spread is commonly measured by three metrics: variance, standard deviation, and range. Range is the difference between the highest and lowest value in the set of data.

The variance, also called the mean square deviation of the values from the average, is given by Equation 2–1.

$$\text{Variance} = S^2 = \frac{\sum_{i=1}^{n}(X_i - \overline{X})^2}{n-1} \tag{2-1}$$

where

X_i is an individual data point
n is the total number of data points
$\overline{X}$ is the average of the data, and is given by Equation 2–2.

$$\overline{X} = \frac{\sum_{i=1}^{n} X_i}{n} \tag{2-2}$$

The standard deviation is the square root of the variance, and is given by Equation 2–3.

$$S = \sqrt{\frac{\sum_{i=1}^{n}(X_i - \overline{X})^2}{n-1}} \tag{2-3}$$

It is not redundant to define both variance and standard deviation. Variance has some useful properties, and so does standard deviation. The significance of S as a metric is depicted in Figure 2–1. If the data are normally distributed, the distance from the average to the point on the curve where the curve changes slope (inflection point) is one standard deviation. About 68% of the data lie between ± S, 95% of the data between ± $2S$, and virtually all of the data (99.7%) are between ± $3S$ [Kittlitz et. al. 1987, Latour 1986]. Thus, the standard deviation is used to estimate how much data falls outside of specifications or limits.

The variability measured in the process is really the total variability, that is, the sum of all the components of variability occurring in the process, including the measurement variability. In order to analyze only the process variability, we really would like to separate the product or process variability from the total variability. Variance can be used to determine the product variability because it has the useful property of additivity. Individual components of variability add up to the total variability. Thus, if we know the measurement variance, we can calculate the product variance (or process variable variance) from Equation 2–4.

$$\text{Product Variance} = \text{Total Variance} - \text{Measurement Variance} \qquad (2\text{–}4)$$

The standard deviation of the product is

$$S_{\text{prod}} = \sqrt{S_{\text{tot}}^2 - S_{\text{meas}}^2} \qquad (2\text{–}5)$$

It would be helpful to know how much measurement variability we can ignore. If S_{meas} is less than 30% of S_{tot}, the effect will be less than 5%. If S_{meas} is less than 10% of S_{tot}, the effect on S_{prod} will be less than 1% and is probably too little to worry about [Juran 1980].

The principle of additivity of variances also enables us to estimate the variance and standard deviation of a dependent variable as a function of one or more independent variables if we know the equation relating them [Kume 1992]. If the equation relating Y and the independent variables X_i is

$$Y = f(X_1, X_2, \ldots, X_n) \qquad (2\text{–}6)$$

the linear relationship between Y and X_i can be estimated by a truncated Taylor series expansion of the equation.

$$Y \sim f(\overline{X}_1, \overline{X}_2, \ldots, \overline{X}_n)$$
$$+ \sum_{i=1}^{n} \frac{\delta f}{\delta X_i} (X_i - \overline{X}_i) \qquad (2\text{–}7)$$

where

$$\overline{X}_i = \text{mean or steady-state value of } X_i$$
$$\frac{\delta f}{\delta X_i} = \text{partial derivative of } f \text{ with respect to } X_i$$

The mean and variance of Y are approximately:

$$\overline{Y} \sim f(\overline{X}_1, \overline{X}_2, \ldots, \overline{X}_n) \qquad (2\text{–}8)$$

$$S_y^2 \sim \sum_{i=1}^{n} \left(\frac{\delta f}{\delta X_i} \right)^2 S_i^2 \qquad (2\text{--}9)$$

where S_i^2 = variance of X_i.

For example, suppose the polymer viscosity V is related to molecular weight W and percent solids P in the polymer by the following equation:

$$V = a\, W^b P^c \qquad (2\text{--}10)$$

The linearized equation for V is

$$V \sim (a\, \overline{W}^b\, \overline{P}^c) + (a\, \overline{W}^b\, c\, \overline{P}^{c-1})\,(P - \overline{P}) \qquad (2\text{--}11)$$
$$+ (a\, \overline{P}^c\, b\, \overline{W}^{b-1})\,(W - \overline{W})$$

The variance S_v^2 of viscosity is approximately

$$S_v^2 \sim (a\, \overline{W}^b\, c\, \overline{P}^{c-1})^2\, S_p^2 + (a\, \overline{P}^c\, b\, \overline{W}^{b-1})^2\, S_w^2 \qquad (2\text{--}12)$$

Therefore, if we know the variance of molecular weight and percent solids, we can estimate the variance of viscosity. This procedure can be applied to determine the variance and standard deviation of inferred variables that either cannot be measured directly or are measured very infrequently, and have to be estimated from a model operating on other measured variables.

EXCEEDING SPECIFICATIONS

One of the benefits of reducing variability is that product property values will be closer to the target, and therefore less will exceed specifications. Assuming for the moment that the data have a normal distribution as in Figure 2–1, we can employ standard statistical techniques to estimate the probability that any data point will fall outside a specification limit [Meyer 1965, Ott 1975]. This can be used to assess current performance and estimate the reduction in off-specification product by reducing variability. Knowledge of the standard deviation and average value are all that we need to estimate the amount of product that exceeds specifications.

To calculate the fraction of data that exceeds a high specification limit X_H, where the average of the data $\overline{X}$, is less than X_H, first calculate Z_H, which expresses the distance between the average and the specification in the number of standard deviations.

$$Z_H = \frac{X_H - \overline{X}}{S} \tag{2-13}$$

Then,

$$\text{Fraction of data} > X_H = 1 - \Phi(Z_H) \tag{2-14}$$

where $\Phi(Z_H)$ is determined from Standard Normal Distribution tables.

To calculate the fraction of data that is less than a low specification limit X_L where the average $\overline{X}$ is greater than X_L, calculate Z_L:

$$Z_L = \frac{X_L - \overline{X}}{S} \tag{2-15}$$

Then,

$$\text{Fraction of data} < X_L = \Phi(Z_L) \tag{2-16}$$

Φ is the Standard Normal Distribution Function and is tabulated in most statistical texts. Although these equations only apply to normally distributed data, if the distribution is fairly bell-shaped and not heavily skewed, they are a fair approximation. An abbreviated table of Φ is given in Table 2–1.

There are a number of statistical techniques to test for normality. One such method is to plot the data on normal probability paper. The paper is scaled so that a normal distribution curve plots as a straight line. If the line is curved, the data

TABLE 2–1　STANDARD NORMAL DISTRIBUTION FUNCTION

z	$\theta(z)$	z	$\theta(z)$
−3.5	.0002	0.0	.5000
−3.0	.0013	0.5	.6915
−2.5	.0062	1.0	.8413
−2.0	.0228	1.5	.9332
−1.5	.0668	2.0	.9772
−1.0	.1587	2.5	.9938
−0.5	.3085	3.0	.9987
0.0	.5000	3.5	.9998

are non-normal. Standard statistical references such as Juran (1980) illustrate this and other techniques.

Example 2–1

Suppose a product has an average impurity concentration of 450 parts per million (ppm), and the standard deviation is 25 ppm. We can estimate how much product exceeds a high specification of 500 ppm using Equations 2–13 and 2–14.

$$\text{Fraction} > 500 \text{ ppm} = 1 - \Phi\left(\frac{500-450}{25}\right) = 1 - \Phi(2) \tag{2-17}$$
$$= 1 - .977$$
$$= .023 \text{ or } 2.3\% \text{ exceeds the specification.}$$

The improvement we can expect for reduced variability can be estimated by comparing Φ for the average operating standard deviation to the improved standard deviation with better control. Some examples will be shown later on.

NON-NORMAL DISTRIBUTIONS

If operating data were normally distributed signifying a state of statistical process control, we could use many of the statistical metrics and tools without reservation. However, in most chemical process operations, there are many sources of variability for which the control system is continuously trying to compensate, and we cannot blindly assume normality.

Real distributions may not be symmetrical but appear skewed to the right or left. One cause of skew is the presence of outliers, which are data that tend to be extreme and erratic. They cause the distribution to have a heavier-than-normal tail. If outliers are present, engineering judgment should be applied to determine if they are real or fluky. Fluky outliers may be discarded to remove some of the skew. Outliers that represent real variability should be investigated. Removing them would result in underestimating the variability.

Another reason for skew may be a characteristic of the measurement instrument. The measurement may have a threshold value above or below which it does not have the sensitivity to detect any changes. The distribution may then have a sharp dropoff. There may also be a one-sided boundary in the process variable. For example, the coating thickness on a sheet of film cannot be less than zero.

Skew can also be caused by characteristics of the process itself. A classic case involves process nonlinearities. A high purity distillation column exhibits this behavior. Suppose the purity of the bottom product is controlled by adjusting boilup. An increase in boilup will not have as great an effect on purity as a decrease in boilup of the same magnitude. The increase in boilup may only change the purity a few ppm, but the same magnitude decrease may cause a reduction in purity of several hundred ppm. This behavior is explained by thermodynamics, which says that it takes proportionally greater amounts of energy to separate components as the fluid becomes more pure. (This is why it is economical to make product that just meets specifications the first time rather than blend underpure with overpure product to meet specifications.)

This nonlinear phenomenon has control implications as well. If the purity gets above setpoint, a given control action will be slower than if the purity goes below setpoint where control action has a greater effect. That is why more higher-purity data are generated than lower-purity data under closed loop control. A way to compensate for the nonlinear behavior is to linearize the process measurement by a control function that has the opposite nonlinear characteristic. For example, if the process nonlinearity is second order, it can be compensated for by a square root function. This is commonly done to compensate for the nonlinear behavior of orifice plates that measure flow. If the nonlinearity is exponential, it can be compensated for by a log function, and so forth.

The distribution of high-purity data tends to be skewed towards higher purity. Pryor (1982) shows a skewed distribution for chlorine residue from a bleaching column that has a heavy tail towards higher percentages of residue. Jacobs (1990) shows the same behavior for oxygen purity. It is probably safe to say that whenever dealing with high-purity materials from a chemical process, assume that the data will be skewed.

A brute-force way in which operating data can be checked for skew is by plotting a histogram of the data and comparing it by eye to a normal distribution curve [Levinson 1990]. The histogram plots values on the x-axis versus the number of occurrences or the percent of the total data points on the y-axis. Instead of plotting values for every data point, it is quicker to group the data into 10 to 30 intervals or blocks of data. For example, if the data range from 0 to 100, divide the data into 10 blocks (0–10, 11–20, . . .) and assign the data to the appropriate blocks based on their value. Plot the intervals (0–10, 11–20, . . .) on the x-axis versus the number of data points in each interval on the y-axis.

Figure 2–2 shows an example of a histogram for a series of purity data.

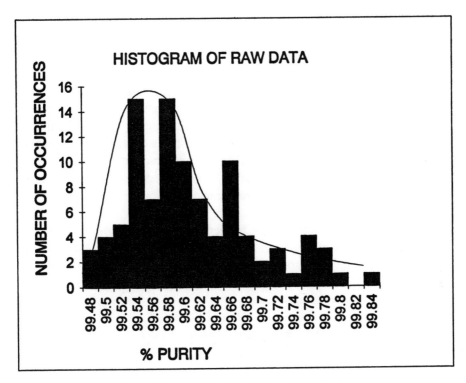

Figure 2–2   An Example of Skewed Distribution

Note that the distribution is not symmetrical about the average but has a heavy tail to the right.

TRANSFORMING DATA

The Standard Normal Distribution function can only be applied to normal data. A common approach for dealing with non-normal data is to transform the data so that the distribution of the transformed data is reasonably close to normal. Then the standard method to calculate the percentage of the data that exceed specifications can be applied.

Data are transformed by applying a single mathematical function to the original data [Hoaglin 1983]. Power transformations commonly have the following general forms:

$$T_1(X) = a X^p + b \quad (p \neq 0) \tag{2-17}$$

$$T_2(X) = c \log X + d \quad (p = 0) \tag{2-18}$$

where $a > 0$ for $p > 0$ and $a < 0$ for $p < 0$, and a, b, c, d are real numbers that can be adjusted to make the transformed distribution fit the normal distribution.

The effect of the log transformation is to compress the scale for the larger data values more than it does for smaller ones. Thus, a distribution that is heavily skewed to the larger values will be made more normal by a log transformation. An X^2 transform compresses the scale for smaller data values more than for larger ones. The point is that the correct transformation depends on the distribution.

Obviously, transformations are needed when the data are heavily skewed. The disadvantage in working with transformed variables is that the scale is less familiar, and we may lose some intuition when dealing with them. Hence, we want to apply transformations judiciously. Hoaglin (1983) offers one rule-of-thumb based on the spread of the ratio:

$$\frac{\text{largest data value} - \text{threshold value}}{\text{smallest data value} - \text{threshold value}}$$

where the threshold value is less than the smallest data value. Transformations should be worthwhile if the ratio is large (> 20) but is probably unnecessary for small ratios (< 2). The more skewed the data or larger the tail, the larger the ratio will be. The threshold value is to take care of a situation where the data are skewed and clustered together. For example, suppose the data range from 99.05 to 99.99. The ratio of 99.99/99.05 is around one. However, using a threshold value of 99.0, the ratio is 20. This is probably a more accurate assessment of the need for transforming the data.

Since the data in Figure 2–2 is skewed to the right, it does not fit a normal distribution, and it would be futile to try to assess how much data exceeds a high specification limit based on the Standard Normal Distribution tables. Figure 2–3 shows the histogram for transformed data using the specific log transform discussed by Jacobs (1990).

$$Y = \ln(X - \Theta), \quad \Theta < X_{\min} \tag{2-19}$$

where Θ is a threshold value. Each original data point was transformed, and then a new histogram was created. Note that the log function compressed the high val-

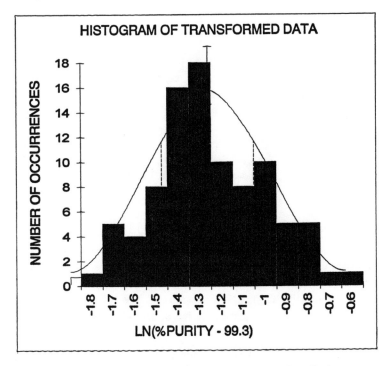

Figure 2–3   Transformed Data to Normalize the Distribution

ues, making the distribution more normal. The transformed distribution can be handled like the normal distribution. Before the Standard Normal Distribution Function can be applied, the specification limits have to be transformed, and the average $\overline{Y}$ and standard deviation S_Y calculated for the transformed X values. Then the transformed Zs are calculated as follows:

$$Z_L = \frac{\ln (X_L - \Theta) - \overline{Y}}{S_Y} \tag{2-20}$$

$$Z_H = \frac{\ln (X_H - \Theta) - \overline{Y}}{S_Y} \tag{2-21}$$

Example 2–2

Find the amount of data below the low limit of 99.5% in Figure 2–3 using the distribution for the transformed data. The average and standard deviation for the transformed data are −1.17 and 0.25, respectively.

From Equation 2–20,

$$Z_L = \frac{\ln(X_L - \Theta) - \overline{Y}}{S_Y}$$
$$= \frac{\ln(X_L - 99.3) + 1.17}{0.25}$$
$$= \frac{-0.44}{0.25} = -1.76$$

Fraction $< 99.5\% = 0.04$ or 4%.

TOTAL VARIABILITY VERSUS CAPABILITY

When the process is under poor control or no control at all, the operating data will contain long-term drifts, shifts, and cycles from special causes plus the inherent random variability from common causes. We refer to the standard deviation that includes variability from all these sources as the *total standard deviation,* S_{tot}, and calculate it by Equation 2–3. Note that throughout the book, it is assumed that the standard deviations reflect the true process variable or product property, that is, the measurement variability is either insignificant (measurement variance is less than 10% of the total variance) or removed from the data.

If statistical and automatic process control are properly applied, most of the extraneous or special-cause variability will be eliminated, and the data will contain only the natural or common-cause variability. This ideal state represents the *capability* of the process. The capability means that this is the least variability that can be achieved with the current process. Any further reduction in variability can only come from some changes to the process itself. Capability is a term used frequently when talking about quality control and ability to meet customer needs. More of this will be covered in the following chapter when quality metrics are discussed.

Figure 2–4 illustrates the concept of total variability and capability. These operating data contain both special-cause and common-cause variability. The total variability is expressed by the histogram on the right side. The spread of data is large, as one would expect under poor control conditions. The short periods of data enclosed in the boxes contain only short-term random variability, so the corresponding histograms are narrow. If the standard deviation for these short periods were calculated, they would reflect the *capability standard deviation,* S_{cap}.

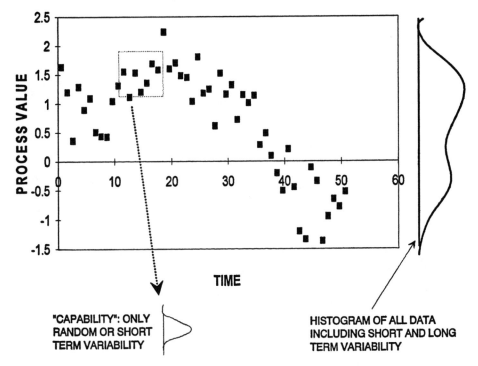

Figure 2–4 Total Variability and Capability

Most of the time, the process will contain more than just random variability, so it may be difficult to measure the capability standard deviation directly without reviewing a lot of data or conducting special plant tests. One way is to review a series of data that represents average operation, and look for those periods of smooth control. These periods are then assumed to describe the capability of the process. An alternate approach is to estimate the capability standard deviation from the average data. This can be done by two methods:

- Average moving range
- Mean square successive difference (MSSD)

The moving range is the absolute value of the difference between successive pairs of data points in a series of operating data [Wadsworth et al. 1986]. The average moving range is the average of the series of moving ranges and is calculated by Equation 2–22.

$$\overline{MR} = \sum_{i=2}^{n} \frac{|X_i - X_{i-1}|}{(n-1)} \tag{2-22}$$

The estimated capability standard deviation from the average moving range is

$$S_{\text{cap}} = \frac{\overline{MR}}{d_2} \tag{2-23}$$

where the constant $d_2 = 1.128$ for pairs of data [Ott 1975, Wadsworth et al. 1986]. By taking differences between successive data points instead of between each data point and the average over the whole population, the long-term variations are cancelled out. Referring again to Figure 2–4, taking differences between pairs of data points is like making the boxes very small.

The estimated capability standard deviation using MSSD is similar to the moving range method. It also operates on the differences between successive pairs of data points [Hald 1952].

$$S_{\text{cap}} = \sqrt{\frac{\sum_{i=2}^{n}(X_i - X_{i-1})^2}{2(n-1)}} \tag{2-24}$$

The results of Equations 2–23 and 2–24 are almost identical.

The estimated capability standard deviation will usually be less than the total standard deviation. However, there are a few cases where the estimates using Equations 2–23 and 2–24 break down. One is when the data have a zigzag pattern, that is, when successive data flip-flop between high and low values. This could be a sign of negative autocorrelation arising when a controller is unstable. When this happens, the sum of the differences between successive data points becomes larger than the sum of the deviations between each data point and the average. Other cases are where the data are random, and where there are outliers in the data. For the majority of cases, though, the capability estimators will work fine.

By comparing the total standard deviation (Equation 2–3) and the capability standard deviation (Equation 2–23 or 2–24), we can estimate the maximum potential improvement in variability with ideal statistical process control and automatic feedback control. Since the focus is primarily on the benefits of automatic process control, the potential reduction in variability for ideal feedback control will be estimated. This is done by comparing the total standard deviation to an estimate of minimum standard deviation under ideal feedback control (i.e., minimum variance control). Note that to achieve minimum variance control, the control action may have to be drastic, and therefore may upset the process too much. However, it does define the variability limits against which current performance can be judged.

The following formula was developed by W.H. Fellner [Bailey 1993] to estimate the standard deviation for minimum variance control.

$$S_{apc} = S_{cap} \sqrt{2 - \left[\frac{S_{cap}}{S_{tot}} \right]^2} \qquad (2\text{--}25)$$

Note that S_{cap} must be limited to less than $\sqrt{2} \, S_{tot}$ to avoid taking the square root of a negative number in Equation 2–25. In theory, S_{cap} should be less than $\sqrt{2} \, S_{tot}$, but this could be violated in a finite amount of data. If $S_{cap} > \sqrt{2} \, S_{tot}$, S_{apc} should be considered zero.

The result of Equation 2–25 compares closely with the method of Harris (1989) for minimum variance under ideal feedback control. Harris uses a time series analysis of process data in which the residual variability during one deadtime is equated to the minimum variance for feedback control. Since most control engineers are not very familiar with time series analysis, Fellner's formula will be applied instead.

Equation 2–26 calculates the percentage of reduction in standard deviation with improved control.

$$\text{Percent Reduction in Standard Deviation} = 100 \left(1 - \frac{S_{apc}}{S_{tot}} \right) \qquad (2\text{--}26)$$

Figure 2–5 is a quick way to estimate the potential reduction in variability with feedback control based on Fellner's formula. It shows how S_{apc}/S_{cap} and the potential to reduce variability change as a function of S_{cap}/S_{tot}. As Scap approaches zero, S_{apc}/S_{cap} approaches the square root of 2, and the potential reduction in variability approaches 100%. As S_{cap} increases and approaches S_{tot}, the process gets more random and S_{apc} approaches S_{cap}. However, the potential ability to reduce variability drops to zero.

If the maximum reduction in standard deviation is not enough, Harris (1989) suggests doing three things:

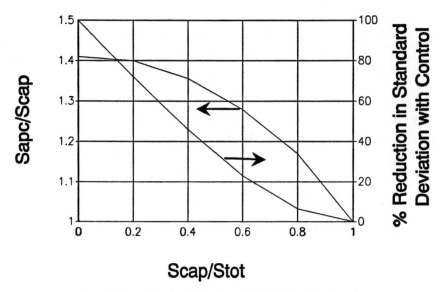

Figure 2–5 Estimating Reduction in Variability from S_{cap}/S_{tot}

- Eliminate deadtime in the process and control loop, including increasing the measurement frequency.
- Reduce the source of upsets by changing the process.
- Change the structure of the control strategy by, for example, using other measurements to apply feedforward control.

COMPENSATING FOR DEADTIME

Equations 2–22 and 2–24 assume that the time delays in the process and control laboratory are short compared to the time between data points (sample period). If this is not true and there are long delays other than the sample period, the equations for S_{cap} should be modified. Instead of taking differences between successive data points, take differences between data that are d units apart [Bailey 1993] where

$$d = \frac{\text{process and lab deadtime}}{\text{sample period}} \qquad (2\text{–}27)$$

The parameter d is rounded off to the nearest whole number. For example, Equation 2–24 would become

$$S_{\text{cap}(d)} = \sqrt{\frac{\sum_{i=2+d}^{n}(X_i - X_{i-1-d})^2}{2(n-1-d)}} \qquad (2\text{–}28)$$

Example 2–3

A reactor is controlled to achieve the desired concentration of a by-product. Figure 2–6 shows how concentration deviated from aim for the same upset condition with a poorly tuned and a well-tuned feedback controller. Control with the poorly tuned controller had a total standard deviation of 1.4 concentration units. The capability standard deviation (S_{cap}) from Equation 2–24 is 0.38 units, and the predicted minimum variability (S_{apc}) with good control, from Equation 2–25, is 0.47. With the well-tuned controller, the total standard deviation was reduced to 0.49, coming very close to the predicted value.

OPERATING CLOSER TO CONSTRAINTS

One of the benefits of reducing variability is to operate closer to process constraints or limits. This is illustrated in Figure 2–7, which shows two distributions of data, one with a wide spread and one with a narrow one. Notice that the average value can be moved closer to the constraint X_L when the spread is narrow, while keeping the amount of material that exceeds the limit the same. The constraint could represent an optimum operating point or perhaps a physical limit beyond which the process is not allowed to operate. The amount by which the average can be moved is given by the following equation [Latour et al. 1986, Martin et al. 1991].

$$\Delta X = \left(1 - \frac{S_{\text{apc}}}{S_{\text{tot}}}\right)(X_L - \overline{X}_{\text{old}}) \qquad (2\text{–}29)$$

The new operating point is

$$\overline{X}_{\text{new}} = \overline{X}_{\text{old}} + \Delta X \qquad (2\text{–}30)$$

Equation 2–29 is strictly true if the data are normally distributed. If the data are skewed, they should be transformed by the methods discussed earlier.

REACTOR COMPOSITION UNDER POOR CONTROL

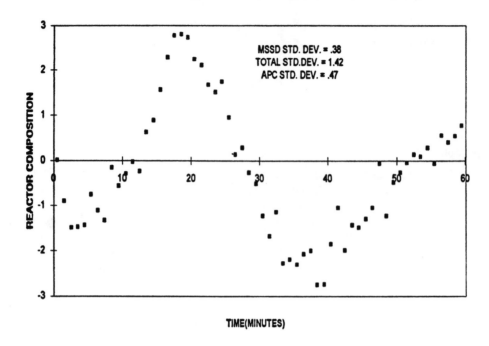

MSSD STD. DEV. = .38
TOTAL STD.DEV. = 1.42
APC STD. DEV. = .47

REACTOR COMPOSITION

TIME(MINUTES)

REACTOR COMPOSITION UNDER GOOD CONTROL

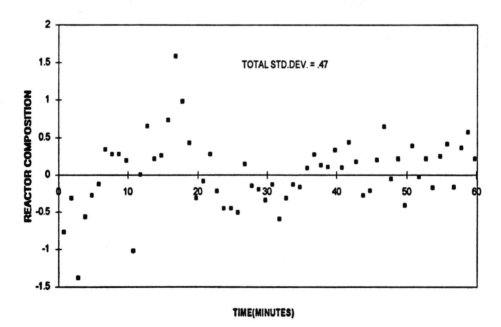

TOTAL STD.DEV. = .47

REACTOR COMPOSITION

TIME(MINUTES)

Figure 2–6 Comparison of Well-Tuned and Poorly Tuned Controllers

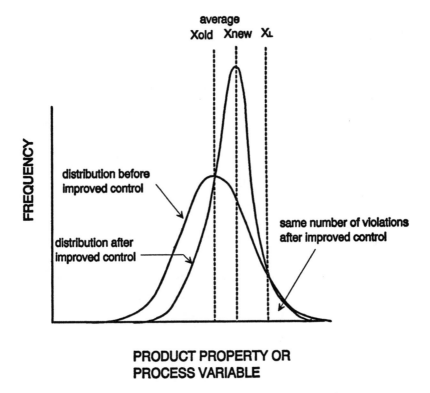

average
Xold Xnew XL

FREQUENCY

distribution before
improved control

distribution after
improved control

same number of violations
after improved control

PRODUCT PROPERTY OR
PROCESS VARIABLE

Figure 2–7   Reducing Variability Allows Operating Closer to Constraints

Example 2–4

A wiped-film evaporator recovers product from a waste stream. Recovering more product has two benefits: it reduces the amount of waste that has to be disposed and increases yield. The amount of product vaporized is controlled by exit temperature. The goal is to control the temperature as close to the high temperature limit of 175 degrees as possible. Exceeding 175 degrees increases the risk of making the waste too viscous, so the operators set the control point conservatively less than 175 degrees because of the wide variability of temperature.

A typical run had an average temperature of 169 degrees and total standard deviation of 5.6 degrees. By improving control, the standard deviation was reduced to 1.3 degrees. Equation 2–29 estimates that the operator could comfortably increase the setpoint 4.6 degrees and not incur any more risk.

$$\Delta T = \left(1 - \frac{1.3}{5.6}\right)(175 - 169) = 4.6 \text{ degrees}$$

TRADITIONAL PROCESS CONTROL
PERFORMANCE CRITERIA

Process control engineers typically do not use the statistical metrics of variance and standard deviation to assess performance of control loops. Instead, they use the integral of squared error (ISE) and integral of absolute error (IAE). These measure the accumulated deviation e between the setpoint and process variable over time.

$$\text{ISE} = \sum_{i=0}^{n} e_i^2 \, Ts \tag{2-31}$$

$$\text{IAE} = \sum_{i=0}^{n} |e_i| \, Ts \tag{2-32}$$

ISE is analogous to variance (Equation 2–1). The number of data points n in the definition of variance equates to t/Ts where t is the total time over which the series of data is collected, and Ts is the sampling period or time interval between data points. Therefore, variance is simply the time average of ISE.

$$\text{Variance} \sim \frac{\text{ISE}}{t} \tag{2-33}$$

ISE is often used as a measure of performance to compare controllers. It is also applied to determine controller tuning by making ISE the objective function in an optimization program that finds the set of tuning parameters to minimize ISE [Deshpande 1988]. It should be noted that minimizing variability may result in very large changes in the manipulative variable, so a controller of this type must be used with caution. Shinskey (1991) shows that the more aggressively a controller is tuned, the less robust it is; that is, the more unstable it becomes if the process conditions and parameters change. IAE is also a popular performance criterion, but, unlike ISE, it weighs the magnitude of errors equally, whereas ISE penalizes larger values of error.

Another measure, accumulated error, (AE), takes the sign of the error into account, so AE is really the net error for a given period of time. It is not used to measure variability because positive and negative errors cancel. Thus, if you had a sustained oscillation, the positive and negative errors would exactly cancel each other, and AE would indicate a variability of zero.

$$AE = \sum_{i=0}^{n} e_i Ts \qquad (2\text{--}34)$$

However, McMillan (1990) points out that AE can estimate the total amount of product that does not meet specifications. By multiplying AE and the average flow rate of the product stream, you can estimate the amount of off-spec product generated during upsets. This has application where liquid streams are eventually to be blended. Thus, although AE is not a good measure of variability, it is useful as an estimator of good product generated during a run.

This book uses the statistical metrics to assess variability for a couple of reasons. Since variance and standard deviation are the most common parameters for measuring and analyzing quality, it makes sense for control engineers to think in those terms also. The other reason is that control engineers and quality control professionals are really both focused on controlling variability, even though their approaches and tools may differ, so it is appropriate to promote understanding and integration of these two approaches to maximize their potential benefits.

REFERENCES

BAILEY, S. P. and FELLNER, W. H. *Some Useful Aids for Understanding and Quantifying Process Control and Improvement Opportunities or How to Deal with the Process Capability "Catch 22."* Paper presented at the ASQC/ASA Fall Technical Conference, Rochester, NY, October 1993.

DESHPANDE, P. B. and ASH, R. H. *Computer Process Control.* Research Triangle Park, N.C.: Instrument Society of America, 1988.

HALD, A. *Statistical Theory with Engineering Applications.* New York: John Wiley, 1952.

HARRIS, T. J. Assessment of Control Loop Performance. *Can. Jnl. of Chem. Eng.,* Vol. 67, pp. 856–861, October 1989.

HOAGLIN, D. C., et al. (eds.) *Understanding Robust and Exploratory Data Analysis.* New York: John Wiley, 1983.

JACOBS, D. C. Watch Out for Nonnormal Distributions. *Chemical Engineering Progress,* pp. 19–27, November 1990.

JURAN, J. M. and GRYNA, F. M., Jr. *Quality Planning and Analysis.* New York: McGraw-Hill, 1980.

KITTLITZ, R., et al. *Quality Assurance for the Chemical and Process Industries.* Amer. Soc. for Qual. Cont., 1987.

KUME, H. *Statistical Methods for Quality Improvement.* Tokyo: AOTS, 1985.

LATOUR, P. L., et al. Estimating Benefits from Advanced Control. *ISA Transactions,* Vol. 25, No. 4, pp. 13–21, 1986.

LEVINSON, W. Understand the Basics of Statistical Process Control. *Chemical Engineering Progress,* pp. 28–37, November 1990.

MARTIN, G. D., et al. Estimating Control Function Benefits. *Hydrocarbon Processing,* pp. 68–73, June 1991.

MCMILLAN, G. K. *Tuning and Control Loop Performance.* Research Triangle Park, N.C.: Instrument Society of America, 1990.

MEYER, P. L. *Introductory Probability and Statistical Applications.* Reading, MA: AddisonWesley, 1965.

OTT, E. R. *Process Quality Control.* New York: McGraw-Hill, 1975.

PRYOR, C. Autocovariance and Power Spectrum Analysis Derive New Information from Process Data. *Control Engineering,* pp. 103–106, October 1992.

SHINSKEY, F. G. Evaluating Feedback Controllers Challenges Users and Vendors. *Control Engineering,* pp. 75–78, September 1991.

WADSWORTH, H. M., JR., et al. *Modern Methods for Quality Control and Improvement.* New York: John Wiley, 1986.

3

Assessing Control Performance

The previous chapter dealt with two fundamental metrics for assessing variability, namely, variance and standard deviation. The metrics are keys to assessing current control performance and developing estimates of how much we can expect to reduce variability through improved control. Chapter 3 applies these metrics to assess how well quality is being controlled. The term *quality* applies to both product properties and the corresponding process variables that need to be properly controlled to achieve product quality.

Quality control is usually concerned with achieving two goals. One is to maintain the quality parameters at some prescribed level, for example, a maximum level of impurity. The other is to maintain consistency in the quality parameters, which is another way of looking at variability.

Variability, as measured by standard deviation, does not by itself address both goals. Clearly, variability is a measure of consistency, but it does not address the goal of maintaining some absolute level or range of the quality parameter. To do this, variability has to be combined with both product specifications (i.e., the minimum and maximum limits of the quality parameter required to meet the customer's needs) and the aim or setpoint.

Chapter 3 describes two metrics that relate specifications and variability: *process performance* indices Pp and Ppk, and *process capability* indices Cp and Cpk. Process capability describes the minimum variability that occurs when the process is in a state of statistical process control, that is, the process variability is only due to the random or common causes inherent in the process. All the special causes of variation have been either eliminated or compensated for by process control. When the capability standard deviation is combined with the customer specifications, we have an index that shows the best quality control possible with the current process.

We can also measure the actual or total standard deviation, which measures variability resulting from all the sources of variability, including extraneous or special causes. When we combine the customer specifications with the total standard deviation, we have a metric called the *process performance index*. In summary, process performance describes the current average operation, while process capability describes the best operation of the current process.

Although the discussion thus far has been on product properties and meeting customer specifications, it applies equally to process variables and meeting standard operating conditions (SOCs) that hopefully are aligned with meeting business drivers.

These two metrics are used to improve quality control by:

- Assessing current performance
- Estimating potential improvement in quality
- Guiding how improvements can be made

PROCESS PERFORMANCE INDEX

The process performance index relates actual product variability to customer specifications. The index is defined by Equation 3–1 [Kittlitz 1987].

$$Pp = \frac{\text{Upper Specification} - \text{Lower Specification}}{6(S_{tot})} \tag{3–1}$$

The histogram in Figure 3–1 helps one to visualize the significance of the index. Virtually all of the data (99.73%) fall between ± 3 standard deviations for a normal distribution. A Pp of 1.0 indicates that the current variability is probably

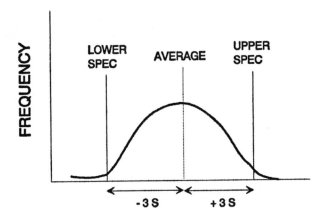

PRODUCT PROPERTY OR PROCESS VARIABLE

Figure 3–1   A Histogram Relates Variability to Customer Specifications

adequate as long as the average of the data coincides with the midpoint of the specification range. Inspection of Figure 3–1 shows that a Pp of 1.0 does not leave much room for error, however.

A variation of the index (Ppk) measures variability relative to the specification range if the average does not coincide with the midpoint of the specification range.

$$Ppk = \frac{|average - nearer\ specification|}{3(S_{tot})} \qquad (3\text{--}2)$$

Therefore, if the average of the data is not at midrange, the performance as measured by Ppk will be less than Pp and will indicate a greater opportunity for improvement. Equation 3–2 is also used when the specification is one-sided.

In summary, then, the process performance index indicates how well the controls are keeping the actual variability within the desired range, and if the product is on-aim. If Ppk equals Pp, the product is on-aim, and if Ppk is less than Pp, it is off-aim. If the specification range happens to equal six standard deviations, and the average of the data corresponds to the midpoint of the specification range, then the data are on-aim and Pp and Ppk are both 1.0. This may be good control for some products, but for others, we may want a smaller variability so that there is some room for error in case we get off-aim. The allowable variability and Ppk really depend on how much we can afford to exceed the specification and still have an acceptable product to sell.

What we attempt to do by improved control is to reduce the standard deviation so that the spread of the data comfortably fits within the specification range and there is room to maneuver and still not exceed the specifications. This point is illustrated by Figure 3–2. If the variability is small compared to the specification range, we can be a little off-aim and still not produce bad product. We would generally prefer to have a Pp of 1.5 or better.

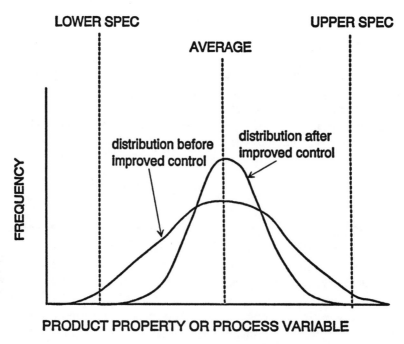

Figure 3–2   Less Variability Means More Room to Maneuver Between Limits

The best performance we can expect from idealized automatic feedback control is given by Equation 3–3, where S_{apc} is defined by Equation 2–25.

$$Pp_{(apc)} = \frac{\text{Upper Specification} - \text{Lower Specification}}{6(S_{apc})} \qquad (3\text{–}3)$$

Note that these equations can also monitor process variables by using SOCs in the numerator.

PROCESS CAPABILITY INDEX

The other common statistical metric relates the minimum possible product variability to the customer specifications. Minimum variability occurs when statistical and automatic process control are being applied perfectly to remove all special-cause variability, leaving only random variability inherent in the process. This metric is called the Process Capability Index [Kittlitz 1987].

$$Cp = \frac{\text{Upper Specification} - \text{Lower Specification}}{6(S_{cap})} \qquad (3\text{–}4)$$

When the process average is accounted for, the index is

$$Cpk = \frac{|\text{average} - \text{nearer specification}|}{3(S_{cap})} \qquad (3\text{–}5)$$

Equation 3–5 is used for one-sided specifications or when the data are off-aim.

Cp represents the smallest variability achievable. It is often used to characterize the process in terms of being able to meet customer requirements. A common goal for Cp is 2.0. In practice, it really depends on what satisfies the customer and the limitations of the particular process. For example, if the customer cannot tolerate any product outside the specification range, the manufacturer should strive for a higher Cp to gain room for errors.

Strictly speaking, there are two requirements for Cp to have any meaning. One is to have the appropriate customer specifications. Second, the standard deviation should be measured when the process is in a state of statistical process control and the data follow a normal distribution [Gunter 1989]. Only then will the total standard deviation equal the capability standard deviation. However, for purposes of estimating process capability for assessing control improvements, we

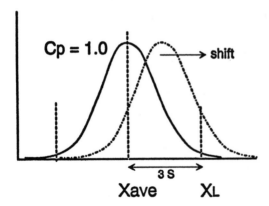

Amount of Shift in standard deviations	% off spec
1 S	2.3%
1.5 S	6.7%
2 S	15.9%

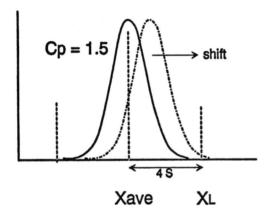

Amount of Shift in standard deviations	% off spec
1 S	0.13%
1.5 S	0.6 %
2 S	2.3 %

Figure 3–3   Comparison of the Percentage of Off-Spec Material Generated by Shifting the Average Value

can approximate capability standard deviation by the average moving range or MSSD.

The importance of controlling variability is illustrated by Figure 3–3, where the percentage of off-spec material generated when the average is off-aim is calculated for two values of Cp. A shift of two standard deviations with a Cp equal to 1.5 is equivalent to a shift of only one standard deviation when Cp equals 1.0.

Cp is like a benchmark for process performance. We would like Ppk to equal Cp if at all possible. That will take a combination of SPC to remove root causes of special-cause variability, APC to compensate for the upsets we cannot feasibly remove, and both to keep the process on-aim.

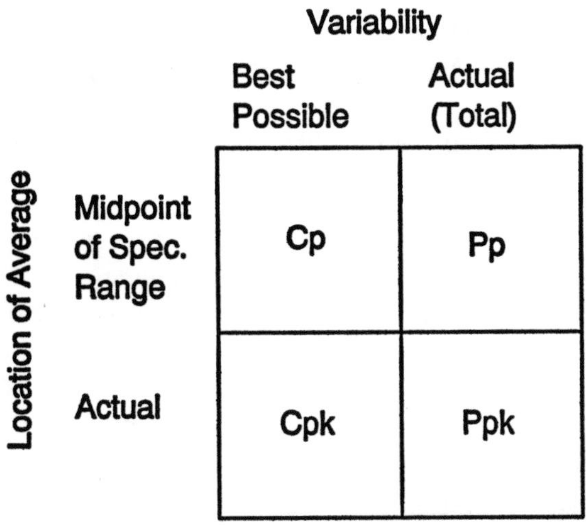

Variability

	Best Possible	Actual (Total)
Midpoint of Spec. Range	Cp	Pp
Actual	Cpk	Ppk

Location of Average

Figure 3–4 Comparison of Performance and Capability Indices

Figure 3–4 illustrates how the various metrics are related.

The best performance is measured by Cp and Cpk, and actual performance by Pp and Ppk. Cp and Pp are used when the process average is at the midpoint of the specification range, and Cpk and Ppk when it is not.

ASSESS OPPORTUNITIES FOR IMPROVED AUTOMATIC PROCESS CONTROL

One reason process control engineers should be interested in applied statistics is that standard statistical metrics are useful tools to help identify process control improvement opportunities. This section shows how performance and capability indices can be applied to do this.

The first step is to analyze the process and identify which product properties are key to satisfying customer needs. This procedure will be examined in Chapter 4. Then apply the performance and capability indices to those properties to assess opportunities for process and control improvements.

COMPARING PPK AND CP

The two questions that should be asked when attempting to improve quality control are:

- Does performance meet capability?
- Does capability meet customer needs?

(In some cases it may be appropriate to consider process variables such as viscosity or concentration and ask if they are conforming to standard operation conditions).

The answers to these two key questions are found by comparing the performance and capability metrics.

- If performance (Ppk) is meeting capability (Cp), there is little incentive to look for process control opportunities. The controls are removing special-cause variability, and that is the best we can expect to achieve by control alone.
- If performance is not meeting capability, there is special-cause variability that may be removed by SPC techniques which search out and eliminate root causes or by compensating for the upsets through APC.
- If capability is meeting customer needs, the process itself warrants no changes.
- If capability is not meeting customer needs, a process change is required.

This analysis is aided by constructing a performance and capability matrix, as shown in Figure 3–5. The indices for each key product property are calculated from average operating data and located in one of the quadrants of the matrix.

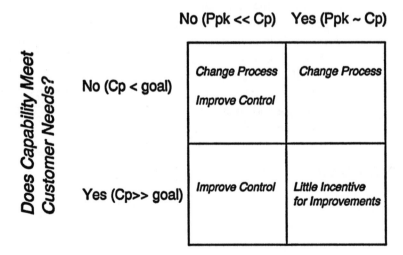

Figure 3–5 Performance and Capability Matrix

Note that Ppk is used to characterize actual performance because it represents the worst case, and Cp is used for capability because it reflects the best case. (For a one-sided specification, however, we substitute Cpk for Cp.) Specific numbers for the indices have been deliberately avoided, because they depend on each situation. However, as rough guidelines, the following numbers can be used [Kittlitz 1987]:

- Cp > 1.5 is considered very good.
- Cp = 1.0 is considered minimally adequate.
- Cp < 0.5 is considered poor.
- Ppk < 0.75 Cp is considered a significant opportunity for improvement.

INTERPRETATION OF THE PERFORMANCE AND CAPABILITY MATRIX

The key to using this tool is interpreting what the various quadrants mean. The following statements interpret the matrix for the case where Cp = 1.0 is considered the minimum adequate goal for capability.

- Cp >> 1 indicates that the process has the capability to meet customer specifications. The inherent random variability that is present is small enough to be acceptable.
- Cp < 1 indicates that the process is not capable of meeting customer specifications, even under statistical process control. There is excessive random variability inherent in the process that must be reduced by a change in the process.
- Ppk << Cp indicates that the process is not operating up to its capability. There are long-term drifts, cycles, shifts, or changes in variability caused by special events that may be reduced by improved process control or eliminated by removing the root causes.
- Ppk ~ Cp indicates that most of the long-term drifts, cycles, and shifts have been removed, and the process is operating close to a state of statistical process control. This is about as good as can be expected from ideal process control.
- Cp < 1 and Ppk << Cp indicate that the random variability is too large, and the process is exhibiting long-term drifts, cycles, and shifts from special causes. Look for ways to eliminate the root causes of both large random

variability and long-term special-cause variability. In addition, compensate with improved APC.

- $Cp < 1$ and $Ppk \sim Cp$ indicate that the random variability is too large, but the special-cause sources of the long-term shifts, drifts, and cycles have been removed. The random variability may be reduced by process changes, but it is unlikely that control can be improved without changing the process.

- $Cp \gg 1$ and $Ppk \ll Cp$ indicate that process control may be improved to remove the special-cause long-term drifts, cycles, and shifts; and/or the root causes for the special-cause variability should be identified and removed, if feasible, by SPC methods.

- $Cp \gg 1$ and $Ppk \sim Cp$ indicate that the process has neither excessive random variability nor long-term drifts, cycles, and shifts. There appears to be little incentive for improvement by either improving control or changing the process.

IMPROVING CONTROL OR CHANGING THE PROCESS

Having determined whether to improve the current control strategy or modify the process, the next step is to consider what specific actions need to be taken to get the improvements. Examples of changing the process and improving the existing controls are given below.

CHANGING THE PROCESS INCLUDES:

- Improving operating procedures
- Changing equipment and/or equipment layout
- Reducing process deadtime
- Optimizing equipment sizes
- Increasing precision or sampling frequency of measurements
- Restructuring pairing of controlled and manipulated variables
- Incorporating other measurements; adding feedforward, multivariable, or other advanced controls

IMPROVING EXISTING CONTROLS INCLUDES:

- Closing control loops
- Optimizing tuning

- Using models to estimate critical properties when measurements are not available

Example 3–1

A dryer removes moisture from a product. Temperature control on the dryer is set to achieve less than 100 ppm water based on a periodic lab analysis. The lab data are shown in Figure 3–6. The statistical metrics are applied to see if the dryer's performance can be improved. A Cp of 1.0 is considered the minimum adequate. The results are shown in Table 3–1.

The dryer controls are as good as we can expect with the current control system because Ppk ~ Cp. However, Cp << 1.0 suggests that the process is not capable of meeting the specification, even with good feedback control. The performance and capability matrix for this example is shown in Figure 3–7, where the upper right quadrant applies.

The matrix suggests that the process must be changed to raise Cp. An investigation showed that the dryer samples picked up moisture during sampling and transportation to the lab, which accounts for the largest component of variability. The problem suggests the need to develop better sample-handling procedures at the

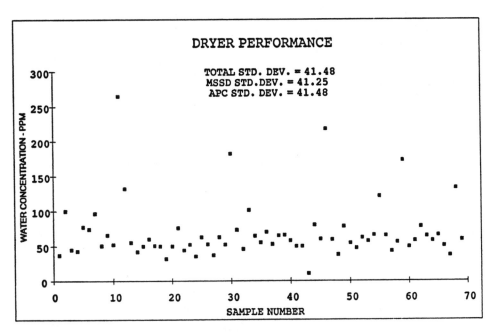

Figure 3–6   Data Showing Dryer Performance

TABLE 3–1   RESULTS FOR EXAMPLE 3–1

Metric	Equation	Value
Average	2–2	70 ppm
S_{tot}	2–3	41.48 ppm
S_{cap}	2–24	41.25 ppm
S_{apc}	2–25	41.47 ppm
Ppk	3–2	.24
Cp	3–4	.242

very minimum, and install an on-line moisture analyzer and tie it into a feedback control loop instead of temperature control.

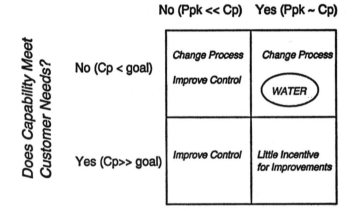

Figure 3–7   Performance and Capability Matrix for Example 3–1

ANALYSIS OF PROCESS VARIABLES

We can extend the quality control metrics Ppk and Cp to measure how the variability of process variables compares to specified operating ranges which are called standard operating conditions (SOCs). Generally, SOCs are set by the manufacturing department. For example, tank level may have an SOC of 25–75% based on what manufacturing thinks is a safe operating range. SOCs are chosen to be some distance away from safety interlocks (e.g., three standard deviations). They may have an alarm so the operator knows when the variable is drifting away from target. All the key process variables that affect quality or other business drivers should have SOCs to optimize the process. *It is important that the SOCs are in alignment with the goals for yield, throughput, quality, and so on.* Otherwise, the controls will not be calibrated to meet those goals.

The Ppk and Cp for process variables are analogous to the quality metrics.

$$\text{Ppk} = \frac{|\text{average} - \text{nearer SOC}|}{3(S_{\text{tot}})} \qquad (3\text{--}6)$$

$$\text{Cp} = \frac{\text{SOC range}}{6(S_{\text{cap}})} \qquad (3\text{--}7)$$

The best Pp or Ppk we can expect from idealized automatic feedback control is given by Equation 3–8.

$$\text{Pp}_{(\text{apc})} = \frac{\text{SOC range}}{6(S_{\text{apc}})} \qquad (3\text{--}8)$$

The performance and capability matrix can be applied in the same way as with product properties to determine how to improve control of the process variables. However, instead of asking if capability meets customer needs, the question would be whether or not capability meets the SOCs' goals. Again, SOCs should be calibrated to meet the goals for yield, throughput rate, quality, or whatever is the key business driver.

The equations above are useful as metrics to track the performance of the key process variables [Kane 1986]. More on performance metrics will be presented in Chapter 9.

THE PATH TO MANUFACTURING IMPROVEMENTS

This chapter and the previous one presented some statistical tools that can identify process control improvement opportunities and estimate the associated stakes. The next two chapters show how the tools are applied.

Figure 3–8 gives a preview. Starting with operating data for the key product properties and process variables—that is, those that have a significant impact on the business drivers—we estimate the three standard deviations: capability, total, and minimum variance. Chapter 4 applies the standard deviations to estimate Cp and Ppk and determine whether a control improvement can reduce variability or

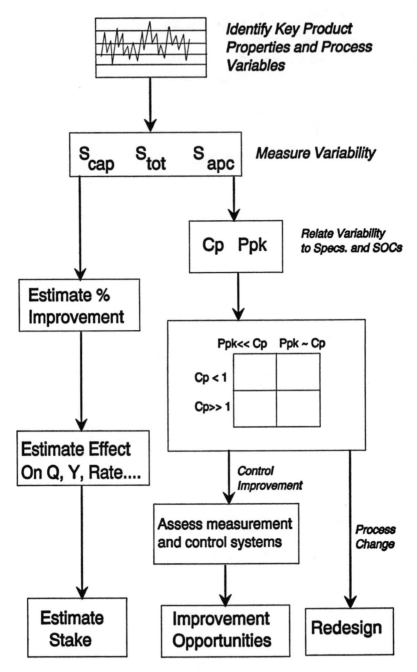

Figure 3–8   The Path to Manufacturing Improvements

the process needs to be modified. If a process control improvement is indicated, the measurement and control systems for those properties and variables are assessed to find the barriers to good control and specific things to improve. If a process change is called for, a redesign of some kind may be warranted.

The standard deviations will be applied in Chapter 5 to estimate the percentage of improvement obtainable through improved control. Examples that show how these improvements affect the business drivers and translate into economic benefits will be presented.

REFERENCES

GUNTER, B. H. The Use and Abuse of Cpk(4 Part Series). *Quality Progress,* January, March, May, July, 1989.

KANE, V. E. Process Capability Indices. *Jnl. of Quality Technology,* Vol. 18, pp. 41–52, January 1986.

KITTLITZ, R., et al. *Quality Assurance for the Chemical and Process Industries.* American Society for Quality Control, 1987.

4

Process Analysis to Identify Control Opportunities

Chapters 2 and 3 introduced statistical metrics that can be applied to product properties and process variables to identify and assess where process control improvement opportunities exist. Since there are hundreds of process variables and many product properties in a typical chemical process, these tools must be applied to those that have a significant impact on the business drivers to achieve any major economic benefits.

Chapter 4 presents a methodology to:

- Analyze the process and determine which product properties and process variables need to be addressed
- Identify opportunities for improved process control

This formal analysis ensures that the cost of making improvements will be based on sound estimates of the benefits.

COST/BENEFIT ANALYSIS

Methods for assessing the value of process control upgrades have been around for a long time. Control consultants often suggest that clients do a cost/benefit analysis as the first step in process control system upgrade projects. This approach obtains significant benefits by focusing on the real issues around improving business performance. The analysis goals are to identify the major control improvement opportunities and estimate the associated costs and benefits. Studies often include a conceptual design or definition of the proposed control upgrades. The study either builds confidence that the money will be well spent or reveals that the project will have marginal value to the business. Once a decision has been made to authorize an improvement project based on the results of an objective and realistic cost/benefit analysis, there is confidence that the money will be well spent.

A study of this type is typically carried out by a small team having the following duties:

- Formulate overall objectives for improving the process.
- Analyze the process and identify where control improvements are needed.
- Generate a list of potential improvement opportunities.
- Estimate the potential benefits and costs for the improvements.
- Prioritize and select the best opportunities based on business impact and feasibility.
- Develop design concepts to capture the benefits.
- Report the results and make recommendations.

Marlin (1987) shows several examples of how these studies have been conducted on various processes. It has a lot of details about who should make up the team, how the interviews should be conducted, and so forth.

This chapter presents a series of steps to analyze the process and identify and assess specific control opportunities to meet the business drivers using the statistical tools introduced in the previous chapters. This analysis produces a list of candidates for improved measurement and control, and estimates the reduction in variability in the key product properties and process variables. It provides the information needed for the cost/benefit estimates and conceptual designs. The steps in this analysis are shown in Figure 4–1.

STEP 1. IDENTIFY THE BUSINESS DRIVERS

For any measurement or control improvement to achieve economic benefits, it must have a significant impact on a business objective or driver. Thus, the first activity is to identify the business drivers. If the process is production-limited, the drivers might be to increase throughput, first-pass first-quality yield, or uptime. If the product is marketlimited, the drivers might be to reduce energy consumption

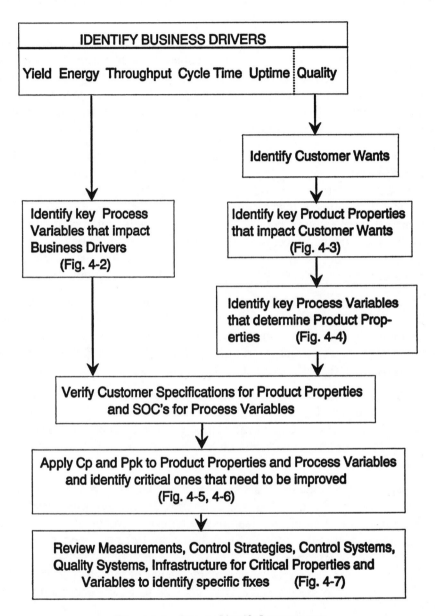

Figure 4–1 Steps to Identify Improvements

and yield losses to reduce costs. Several common business drivers are shown at the top of Figure 4–1.

• *Yield* measures the percentage of raw materials that are converted to useful products. Yield improvements can be gained by reducing physical losses of materials in vents and waste streams, and by reducing chemical reactions in which undesirable by-products are generated. *First-pass first-quality yield* measures the percentage of raw materials that are converted to product that meets the desired specifications without blending, rework, recycling, or price concessions.

• *Throughput* refers to the rate at which product is made. This business driver is enhanced by removing any bottlenecks in the process. Low first-pass first-quality yield also impacts throughput when off-spec material is reprocessed, which ties up equipment that could be producing new product instead. Bottlenecks are often the result of poor control and can be eliminated by reducing variability. This allows the process to be operated closer to operating constraints or limits. If the product is sold out, an increase in throughput translates directly into increased earnings, which usually makes it a very strong economic driver.

• *Cycle time* is the elapsed time from when raw materials are received until product is shipped. Control improvements that reduce variability can play a big part in reducing in-process storage and improving equipment reliability and uptime, which ultimately affect cycle time. In batch processes, throughput may be increased by reducing batch cycle time.

• *Uptime* is the percentage of time the process can run at full rate and make first-quality product. Uptime can be adversely affected by unscheduled shutdowns to clean out fouled or plugged equipment. Fouling and pluggage are often caused by poor control. This occurs quite frequently when process streams contain solids.

• *Energy* is the amount of steam, fuel, and electricity that are consumed to make a unit of product. Energy savings in terms of steam consumption is easily related to improved control. For example, in a distillation column where process streams are purified, poor control of variability can force the operator to run at a much higher steam rate than required to achieve the same purity with good control. Energy control also refers to optimizing the use of refrigeration systems and steam boilers. Energy consumption is reduced as first-pass first-quality yield is improved due to less rework.

• *Quality* is making product that fits customer needs. The path to quality is the reduction of variation in the product properties and process variables that affect quality. Quality may be concerned with consistency (i.e., making product the same way all the time) and/or the absolute levels of impurities.

STEP 2. LINK KEY PRODUCT PROPERTIES AND PROCESS VARIABLES TO BUSINESS DRIVERS

There are usually a relatively small number of key process variables that have a major impact on the business drivers compared to the total number, which may be in the hundreds. On the other hand, most of the product properties that are measured have a significant impact on quality. In general, product property analyses are relatively expensive and should be pared down to only those essential to ensure that product will meet customer needs. In either case, these key product properties and process variables must be the focus of any process control improvements to really achieve significant economic benefits.

When considering yield, throughput, cycle time, energy, and uptime, process variables are usually the focus for control improvements. For example, if a distillation column is the bottleneck in achieving the desired throughput, a key variable might be column pressure. At higher pressures, the vapor is denser and the column capacity in terms of mass flow of vapor increases. Notwithstanding improvements in relative volatility at lower pressures, for many separations this may be the overriding factor in improving capacity. In another example, if a heat exchanger fouls frequently, causing reduced rates, a key variable might be the exit process temperature.

The process of linking the business drivers to the key process variables can be conducted by a simple process and aided by Figure 4–2, except for quality. Quality will be handled in a different manner. Process variables that are believed to be significant in meeting one or more business drivers are listed at the top. Each business driver is weighted (1–5) to reflect relative importance. The impact of each process variable on each driver is rated on a scale of 1–5 (the higher number implies a greater impact) and entered in the appropriate box in the matrix. The weighting factor for each driver is multiplied by the rating for each variable, and the result is entered in the shaded box. Then the numbers for each variable are totalled to get their relative importance. The ones having the highest scores become candidates for improved control.

Obviously, the team carrying out the analysis needs to have process experts on it. A lack of process understanding may require that plant experiments or sim-

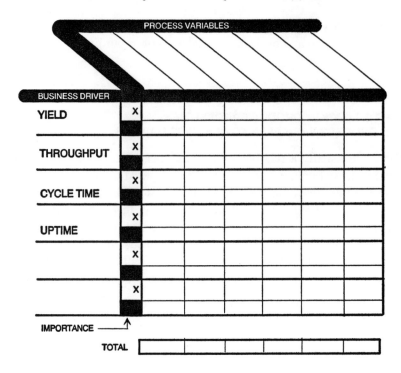

Figure 4–2   Linking Business Drivers and Process Variables

ulations first be conducted to establish the relationships between the business drivers and the process variables.

Sometimes it is difficult to list key process variables that affect the business drivers right off the top of one's head. If that is the case, a way to get started is to have the top operators list what process variables they tend to watch more closely for each equipment piece. Then use that list as a starting point for finding the variables that affect the business drivers.

In contrast, the product quality driver should focus on the product properties to make sure they conform to the customer specifications. Thus, the first thing to do is identify the customer requirements and then link them to the key product properties. Figure 4–3 is a worksheet for finding the key properties for quality control. The procedure is analogous to Figure 4–2 and also assumes that the appropriate level of process expertise is available. It is often necessary to con-

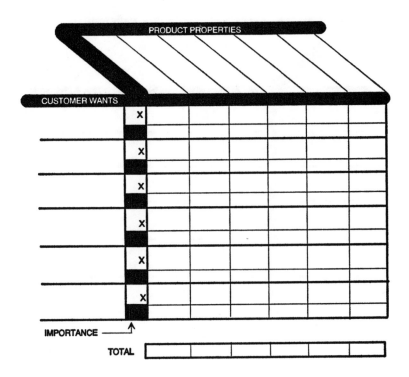

INSTRUCTIONS

1. List customer wants . Prioritize by a weighting factor (1-5)
2. List product properties.
3. Enter rating of each product property according to impact on customer want (1-5)
4. Multiple weighting x rating and add columns to get total rating

Figure 4–3   Linking Customer Wants and Product Properties for Quality Control

duct interviews with the customers directly to get information on what their needs are.

Since a process variable is ultimately adjusted to control a product property, those key variables must also be identified. Figure 4–4 is a cause-and-effect analysis that links the appropriate process variables to the product properties. The

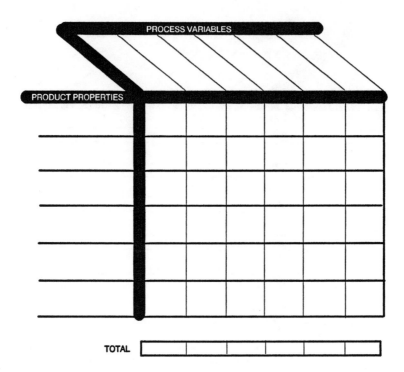

Figure 4–4   Linking Product Properties and Process Variables for Quality Control

key properties from Figure 4–3 are listed on the left side. Along the top, the variables that are known to affect at least one of the properties are listed. Each variable is rated according to its relationship to the properties (1–9). The key variables that have a major impact on the key properties are the candidates for improved control.

The metrics Ppk and Cp are applied to the results of these steps to evaluate performance and capability of the process and determine which are in need of control improvements. This is described in Step 4.

STEP 3. VERIFY CUSTOMER SPECIFICATIONS FOR PRODUCT PROPERTIES AND SOCS FOR PROCESS VARIABLES

Obviously, having valid customer specifications and SOCs is necessary for finding the best opportunities to achieve the business goals. Determining customer specifications for each product property is not always an easy task. Kittlitz (1987) suggests some ways to do this. Some of the suggestions have been paraphrased below.

• Whenever possible, state specifications in terms of true product properties, excluding measurement variability. This implies that the measurement variability will have to be assessed and, if too large, separated from the total variability (see Chapter 2).

• Base specifications on customer requirements. This demands that you know the customer requirements. Very often, a supplier has only a vague idea of customer needs, and therefore has no valid basis for setting specifications.

• If customer requirements are unknown, the supplier and customer should jointly develop and agree upon the specifications.

• The starting point for developing specifications should be the supplier's demonstrated process performance and the customer's experience.

• If the customer's experience has been satisfactory with the product, specifications should be set close to the supplier's demonstrated performance.

• If the customer's experience has not been satisfactory, specifications should be agreed on that the supplier can meet and the customer can accept.

SOCs for process variables are often based on operational objectives rather than business objectives. For example, SOCs may be arbitrarily set a certain distance from process interlocks. For the key variables, we want the SOCs to be based on achieving the business goals. For example, if throughput is maximized by operating equipment within a certain pressure range, we should base the SOC on that criterion. An SOC for temperature control in a distillation column may be set on the basis of controlling an impurity in the product.

It is not always easy to relate SOCs to the business drivers, but considering the potential economic benefits, it may be worthwhile to do it. The task may involve making special plant tests, modeling and simulation, or developing correlations from historical data. The task may be complicated by the multivariable, interactive, and nonlinear behavior of the process. Some advanced modeling tools, such as partial least squares and principal component analysis, may be required in these cases [MacGregor et al. 1991].

STEP 4. APPLY Cp AND Ppk

Step 4. quantifies the variability of the key product properties and process variables to determine which ones need to be improved. These critical ones are the final candidates for improved control. The metrics Cp and Ppk, presented in Chapters 2 and 3, are applied to make the assessment. Figure 4–5 can be used to gather the necessary data and construct the performance and capability matrix. The properties and variables in the upper left quadrant are the most critical in terms of needing improvement, the next critical the upper right quadrant, then the lower left quadrant.

We can go one step further and assess the potential reduction in variability for improved feedback control using the formulas in Chapters 2 and 3. Figure 4–6 aids in that task. These numbers may be used for estimating the economic benefits of the improvements.

STEP 5. REVIEW MEASUREMENTS, CONTROL STRATEGIES, QUALITY AND CONTROL SYSTEMS, AND PLANT INFRASTRUCTURE FOR THE CRITICAL PROPERTIES AND VARIABLES

The final step in the analysis is to assess the systems and support infrastructure for the critical properties and variables to identify the barriers to achieving the desired performance. These represent specific opportunities that can hopefully be fixed to achieve the business goals. Figure 4–7 is an aid in this task. It lists the kinds of questions that should be asked in assessing the performance of the systems. The specific barriers to achieving the business goals are captured as potential opportunities for improvement.

KEY PRODUCT PROPERTY	Cp goal	S_{tot}	S_{cap}	AVE	UPPER SPEC	LOWER SPEC	Cp	Ppk

KEY PROCESS VARIABLE	Cp goal	S_{tot}	S_{cap}	AVE	UPPER SOC	LOWER SOC	Cp	Ppk

	Ppk << Cp	Ppk ~ Cp
Cp < goal		
Cp >> goal		

Figure 4–5 Process Performance and Capability Worksheet

KEY PRODUCT PROPERTY	S_{tot}	S_{cap}	S_{apc}	% red in S	UPPER SPEC	LOWER SPEC	new Ppk

KEY PROCESS VARIABLE	S_{tot}	S_{cap}	S_{apc}	% red in S	UPPER SOC	LOWER SOC	new Ppk

PARAMETER	FORMULA
Stot	2-3
Scap	2-23 or 2-24
Sapc	2-25
% reduction in S	2-26
new Ppk (product property)	3-3
new Ppk (process variable)	3-8

Figure 4–6   Estimate of Improved Variability

Process [   ]   **Critical Property or Variable** [   ]

Category	Potential Opportunities for Improvement (x) if a barrier to meeting business goal		Identify Opportunity
Measurement	Accuracy, precision, sensitivity Repeatibility Reliability/uptime Frequency Time delay/turnaround time Appropriate SOC/specification Measurement/analysis technique Sample procedures Laboratory practices Lack of measurement	(x)	
Control Strategies	Operates in intended mode Handles upsets/stable Tuning adequate Loop interactions Adequate turndown capability Stable over wide operating range Appropriate for control task Complexity Control strategy /systems support		
Quality Systems	Appropriate control charts Appropriate alms, limits Data handling(outliers, normality, correlated) Root-cause elimination procedures Sampling frequency Sample size Corrective action Appropriate variables		
Infrastructure/ Systems Support	Management support for control Prioritized technical programs Adequate process understanding Enough manpower Adequate training of operators, engineers, technicians Preventive maintenance practices Controller tuning practices Clearly defined responsibilities Linking to other technology centers		

Figure 4–7   Process Systems Assessment

Through the process just outlined, the number of opportunities is reduced to the significant ones that need to be addressed. The list of opportunities can be further processed to estimate capital and non-capital costs, and to develop a conceptual design for the improvements.

Example 4–1

The process shown in Figure 4–8 makes several grades of a polymer. The important business driver is quality. The customer wants the product properties to be consistent and have a low water content and low color, and a fast response to orders for different grades. The main properties that affect the customer are molecular weight distribution, percentage of water, and percentage of ash. Molecular weight is measured in the control lab and is indirectly controlled by adjusting the reactor feed ratio. Reactor residence time and temperature also affect molecular weight. Poor molecular weight control causes the polymer to be inconsistent as well as long transition times between grade changes, and affects cycle time. If the filter system is not operating properly, the catalyst shows up as ash in the polymer and causes color problems. Poor dryer control causes excess water in the polymer.

This example shows how the analysis is done to pinpoint the major opportunities for improved control. The analysis is illustrated in Figure 4–9.

- Figure 4–9A relates customer wants to product properties. Molecular weight has the largest impact on the customer.
- Figure 4–9B relates the key process variables to molecular weight by a cause-and-effect analysis. Reactor feed ratio is believed to have the greatest impact on molecular weight.
- Figure 4–9C shows Cp and Ppk. Cp's for molecular weight and feed ratio both fall short of the goal of 1.5, but Cp for molecular weight is very poor by comparison. There is a large gap between Cp and Ppk for feed ratio, but the gap is not as great for molecular weight. In other words, there is a great opportunity to reduce variability through control with feed ratio, but not as much for molecular weight.
- The Cp vs. Ppk matrix (Figure 4–9D) indicates that control improvements and process changes will be needed to improve the variability of feed ratio, but improving molecular weight variability will mainly require process changes. Molecular weight variability will obviously be helped by reducing the variability of feed ratio.
- Figure 4–9E estimates the improvement potential in Ppk with improved control. There will be a large improvement in variability for feed ratio (72%) and for molecular weight; the percentage of improvement is much less (34%). Process changes are required to get an acceptable variability.

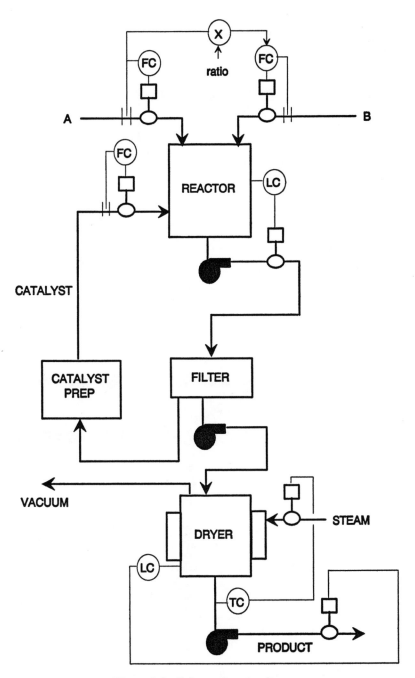

Figure 4–8 Polymer Reaction System

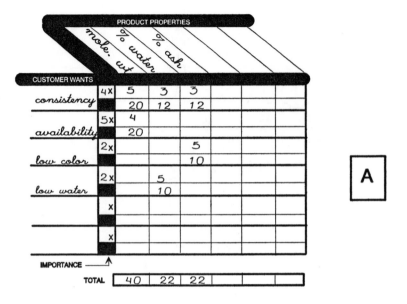

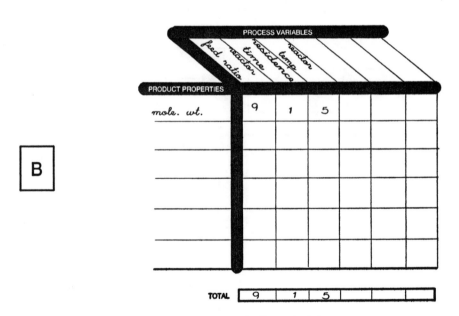

Figure 4–9   Worksheets for Example 4–1

KEY PRODUCT PROPERTY	Cp goal	S tot	S cap	AVE	UPPER SPEC	LOWER SPEC	Cp	Ppk
molecular weight	1.5	50	25	500	525	475	.33	.17

C

KEY PROCESS VARIABLE	Cp goal	S tot	S cap	AVE	UPPER SPEC	LOWER SPEC	Cp	Ppk
reactor feed ratio	1.5	.5	.1	1.2	1.5	1.0	.8	.16

D

	Ppk << Cp	Ppk ~ Cp
Cp < goal	reactor feed ratio	mole. wt.
Cp >> goal		

E

KEY PRODUCT PROPERTY	S tot	S cap	S apc	% red in S	UPPER SPEC	LOWER SPEC	new Ppk
molecular weight	50	25	33	34	525	475	.25

KEY PROCESS VARIABLE	S tot	S cap	S apc	% red in S	UPPER SPEC	LOWER SPEC	new Ppk
reactor feed ratio	.5	.1	.14	72	1.5	1	.6

Figure 4–9 (Continued)

The following actions are indicated:

- The feed ratio controller should be checked to see if it operates well in automatic (potential control improvement).
- The feed measurements should be checked for accuracy, sensitivity, and precision. New measurement instrumentation may be needed (process change).
- The raw material composition variability should be checked. It may be a source of variability (process change).
- The sampling and analytical procedures and methods should be checked for molecular weight. These may be a source of variability and may be alleviated by an on-line measurement (process change).
- Startup and transitioning procedures should be reviewed to see how they may be affecting variability.

REFERENCES

DYBECK, M. and BOZENHARDT, H. Justifying Automation Projects. *Chemical Engineering,* pp. 113–116, June 20, 1988.

MACGREGOR, J. F., et al. *Multivariable Statistical Methods in Process Analysis and Control,* presented at CPC-IV Conference, South Padre Island, TX, February 18, 1991.

MARLIN, T. E., et al. *Advanced Process Control Applications—Warren Centre Industrial Case Studies of Opportunities and Benefits.* Research Triangle Park, NC: Instrument Society of America, 1987.

5

Estimating Benefits

Chapter 4 introduced a series of steps to identify the critical product properties and process variables that need to be improved. If we were doing an analysis of a process to determine how to improve manufacturing by improved control, we would focus on those as having the greatest potential benefits.

Estimating the benefits for improved control is strongly recommended whenever control improvement projects are planned. Knowledge of the benefits provides a basis for justifying capital expenditures. The benefits also indicate how and in what order the improvements should be implemented if there are limited funds available. A great intangible benefit is that it builds confidence and energizes people because they know their efforts will be rewarded by a better-performing plant. Finally, having an idea of how the plant should perform provides a benchmark to assess performance now and in the future.

This chapter presents some examples of how benefits are estimated using the equations presented in Chapters 2 and 3.

YIELD IMPROVEMENTS

A process yield improvement can be gained by:

- Reducing physical losses of materials in vents and waste streams
- Reducing chemical losses through chemical reaction

An example of a physical yield loss is material escaping from a condenser vent. This can happen frequently when process capacity is pushed to the limits. Chemical yield losses occur in reactors when the operating conditions favor formation of waste byproducts. A typical situation in which both chemical and physical losses occur is a reactor-distillation system where an excess of one reactant is fed to keep reactor conversion low and chemical yield high. The excess reactant is recovered from the product in the distillation column and recycled to the reactor. This introduces a physical loss of reactant in the condenser vent. This is a classic case of the need to optimize operating conditions to minimize the total cost of physical and chemical losses.

Physical yield improvements sometimes have to be balanced against higher overall energy costs. For example, more product can be recovered in an absorption column by increasing absorbent flow, but more energy will be required downstream to separate the product from the absorbent.

First-pass first-quality yield is another way to look at yield. This refers to making the product right the first time. The yield may be the result of various processing errors along the way, not necessarily chemical reaction. When first-pass first-quality yield is less than 100%, the off-grade product may have to be reworked or recycled. In a sold-out condition, this ties up equipment that otherwise could be making new product.

Equation 5–1 is a generic equation for calculating the reduction in physical yield losses.

$$Y1 = \left(\text{Flow rate } \frac{\text{units}}{\text{hr}} \right) \times \left(\frac{\text{\% component lost}}{100\%} \right)$$
$$\times \left(\frac{\text{\% reduction}}{100\%} \right) \times \left(\frac{\$}{\text{unit}} \right) \times \left(\frac{\text{hr}}{\text{yr}} \right) \tag{5–1}$$

The reduction in raw material loss due to chemical yield can be calculated by Equation 5–2.

$$Y2 = \left(\text{Raw material flow rate } \frac{\text{units}}{\text{hr}} \right)$$
$$\times \left(\frac{\text{\% recoverable yield loss}}{100\%} \right) \times \left(\frac{\text{\% reduction}}{100\%} \right)$$
$$\times \left(\frac{\$}{\text{unit}} \right) \times \left(\frac{\text{hr}}{\text{yr}} \right) \tag{5–2}$$

Recoverable yield loss is the difference between the current average yield and the theoretical maximum yield for the raw material being considered. Reduction in yield losses is usually achieved by improving the measurement of the controlled variable (e.g., reactor temperature) and minimizing the variability of the feeds and reactor conditions.

In some cases, loss of raw material to the primary product is partly offset by forming a by-product that has some net worth. If so, the cost of the raw material will have to be reduced to reflect this.

Example 5–1

This example is a reactor in which a catalyst is used to enhance the chemical yield. Catalyst concentration should be precisely controlled to maximize yield. Plant data indicate that maximum yield occurs at maximum catalyst concentration; however, catalyst concentration is limited to 0.55% to avoid pluggage. Pluggage causes process downtime, which is costly in a sold-out market. A linear regression equation relates yield and catalyst concentration.

$$\text{Yield} = 80\% + 9 \ (\% \ \text{Catalyst})$$

Figure 5–1 shows how catalyst concentration varies over time. The average concentration is 0.415%, and the total standard deviation is 0.12%. The minimum standard deviation with improved control, calculated from Equation 2–25, is 0.08%. The histogram shows that by reducing the standard deviation, the average concentration can be moved 0.045% closer to the limit, based on Equation 2–29.

$$\Delta \ \% \ \text{catalyst} = \left(1 - \frac{S_{\text{apc}}}{S_{\text{tot}}}\right)(X_L - \overline{X}_{\text{old}})$$

$$= \left(1 - \frac{.08}{.12}\right)(.55 - .415 = 0.045\%$$

This results in a yield improvement of 0.4%.

$$\text{Yield Improvement} = 9 \ (0.045\%) = 0.4\%$$

THROUGHPUT IMPROVEMENTS

Throughput or capacity improvements may offer tremendous benefits in a sold-out market. If the product is sold out, the focus should first be on throughput improvements. Also, consideration should be given to first-pass first-quality yield, because if the off-grade product must be reworked, it may tie up equipment. Improved control can often increase the production capacity of a process by:

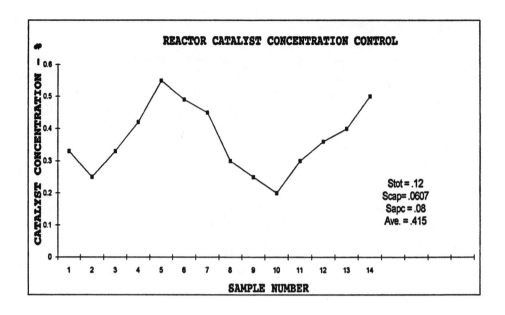

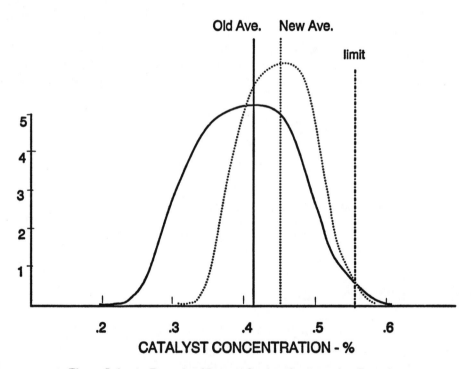

Figure 5–1 An Example of Reactor Catalyst Concentration Control

- Removing production bottlenecks
- Allowing the process to operate closer to constraints
- Reducing downtime due to plugging or corrosion caused by not maintaining the proper operating conditions

In batch processes, throughput can be increased by reducing batch cycle times. If the product is sold out and the process is running at maximum rates, an increase in throughput translates into increased earnings. This is one situation in which process control can gain dramatic benefits.

The value of increased production is calculated by Equation 5–3.

$$P = \text{Current annual production rate}$$
$$\times \left(\frac{\% \text{ estimated increase in throughput}}{100\%} \right) \tag{5–3}$$
$$\times \left(\frac{\% \text{ time sold out}}{100\%} \right) \times \left(\frac{\$}{\text{unit of increased production}} \right)$$

The percent of time sold out is the percent of time the plant needs to run at maximum throughput to meet demand.

If the increased quantity of product can be sold, the value is the incremental price per unit. If product is purchased from an outside vendor to meet sales commitments, the value is the difference between the unit cost to purchase the product and the incremental unit cost of manufacturing.

Example 5–2

Figure 5–2 shows a wiped-film evaporator that recovers product from a waste stream. Recovering more product has the following benefits:

- Increases the throughput of the process, which increases earnings in a sold-out market
- Reduces the amount of waste that has to be incinerated, which is a direct cost savings

In this example, there is a high temperature limit of 175 degrees, which cannot be exceeded for very long without the fluid becoming too viscous to flow. In order to maximize recovery and profitability, the evaporator should be operated as close to the high temperature limit as possible.

The performance was monitored to determine the average and the best standard deviation (S), as shown in Table 5–1.

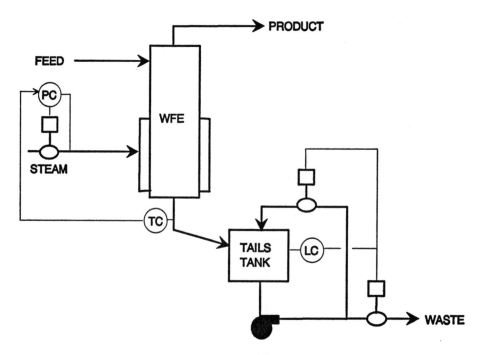

Figure 5–2   A Wiped Film Evaporator System

The relationship between the percentage of product in the waste stream and temperature is estimated from the data:

$$\frac{\% \text{ Product}}{T} = -0.43 \ \%/C$$

Equation 2–29 determines how much closer to the optimum temperature the setpoint can be moved based on the reduction in standard deviation.

$$\Delta T = \left(1 - \frac{S_{apc}}{S_{tot}} \right) (T_L - \overline{T}_{old})$$

$$= (1 - 1.3/5.6) \ (175 - 169) = 4.6 \ C$$

TABLE 5–1   PERFORMANCE FOR EXAMPLE 5–2

Operating Performance	Average Temp./S	% Product in Waste/S
Average	169/5.6	7.75/1.81
Best	171/1.3	7.06/.37

The new setpoint is

$$\overline{T}_{new} = \overline{T}_{old} + \Delta T = 169 + 4.6 = 173.6 \ C$$

The reduction in waste product is

$$\Delta\% = (\Delta T)\left(\frac{\% \ Prod}{T}\right) = 1.98\%$$

The reduction in the percentage of product in the waste stream is

$$\% \ Product = 7.75 - 1.98 = 5.77\%$$

which is a 25% improvement.

QUALITY IMPROVEMENTS

Controlling quality is a fundamental requirement for any manufacturing operation. Quality obviously impacts sales volume and market share. Poor quality affects customers directly, causing complaints and rejected products. Furthermore,
quality affects many other aspects of manufacturing, such as yield, throughput,
waste, inventories, technical support costs, and lab testing costs.

It is fairly easy to relate quality problems to manufacturing problems. However, it is often difficult to relate quality problems to sales volume, market share,
or selling price. A generic equation for estimating the benefits for quality in terms
of increased earnings is given below.

$$Q = \left(annual \ production \ rate \ \frac{units \ of \ product}{yr}\right)$$
$$\times \left(\frac{\% \ increase \ in \ earnings \ for \ higher \ quality}{100\%}\right) \quad (5\text{--}4)$$
$$\times \left(\frac{\$ \ earnings}{unit \ of \ product}\right)$$

Example 5–3

A crude product is purified by distillation. The specification for the product is that
impurity be less than 565 parts per million (ppm). If product impurity is greater
than 565 ppm, it has to be recycled. Figure 5–3 shows the concentration over a period of time. The total standard deviation is 26.3 ppm. Equation 2–25 estimates a
minimum standard deviation of 22.6 ppm with improved control. Reducing the

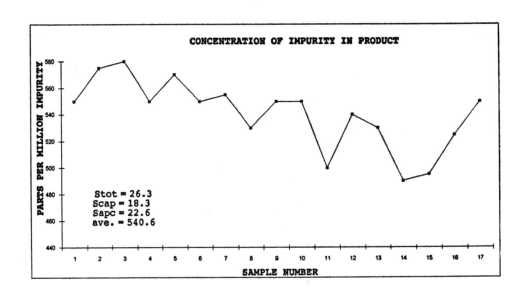

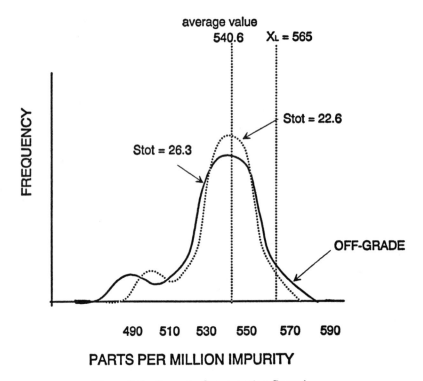

Figure 5–3 Impurity Concentration Control

standard deviation will narrow the distribution, so the percentage of off-grade product will be reduced according to the histogram. We can estimate the reduction in off-grade product with Equation 2–14 and Table 2–1.

The fraction of off-grade product with the total standard deviation of 26.3 ppm

$$= 1 - \Phi\left(\frac{X_H - \overline{X}}{S_{tot}}\right)$$

$$= 1 - \Phi\left(\frac{565 - 540}{26.3}\right)$$

$$= 1 - \Phi(0.95) = 0.17 \text{ or } 17\%$$

The fraction of off-grade product with the improved standard deviation of 22.6 ppm

$$= 1 - \Phi\left(\frac{X_H - \overline{X}}{S_{apc}}\right)$$

$$= 1 - \Phi\left(\frac{565 - 540}{22.6}\right) = 1 - \Phi(1.12) = 0.13 \text{ or } 13\%$$

Thus, the percentage of off-grade material and rework is reduced by 4% for a 24% improvement.

This example clearly shows the impact of better quality control in a sold-out market. A 4% increase in throughput is gained because the plant does not have to rework as much off-grade product. Another benefit is a reduction in off-grade product inventory. Third, there is often an opportunity to reduce the frequency of control lab analyses on final product by reducing variability and raising Ppk through better control.

ENERGY CONSUMPTION IMPROVEMENTS

Energy savings can come from steam, refrigeration, electrical power, and fuel consumption. Energy savings in steam consumption is usually easily related to improved control. About 40% of the energy used in a wet chemical process is for distillation columns to separate materials and refine products. Poor control of product quality causes the operator to overpurify product so that impure product generated during upsets can be blended with the overpure material. The downside is that it takes a greater proportion of energy to achieve an incremental increase in purity as the level of purity increases. Improved control should:

- Reduce variability so that the operator can comfortably reduce the purity target to be closer to the specification
- Provide a more accurate measure of quality

Energy consumption can often be reduced by optimizing the energy level. For example, in some refrigeration systems it is possible to reduce costs by controlling coolant temperature as high as possible, as indicated when one of the user's coolant valves is wide open. Obviously, the coolant is too cold when every user's valve is throttled way back.

Fuel costs for generating steam can be reduced by combustion trim controls and allocating loads to boilers to achieve maximum overall efficiency.

Energy savings are calculated by Equation 5–5.

$$\text{Energy} = \left(\text{energy usage rate } \frac{\text{units}}{\text{hr}}\right) \times \left(\frac{\% \text{ reduction}}{100\%}\right)$$
$$\times \left(\frac{\text{variable cost}}{\text{unit}}\right) \times \left(\frac{\text{hours of production}}{\text{yr}}\right) \quad (2-6)$$

The variable cost is that portion of the total energy cost which will decrease proportionally with a drop in consumption.

Example 5–4

HCN gas is absorbed from a vapor stream in an absorber using recirculating water, as shown in Figure 5–4. The absorber tails feeds a stripper column to separate the absorbed HCN from the water. The water is recirculated back to the top of the absorber after it is cooled. The water rate is fixed at a high level regardless of feed to the absorber. When absorber feed rates are low, as in this case, the excess water absorbs more HCN than is required, puts an unnecessarily high load on the stripper, and thus wastes steam. The average concentration limit of HCN in the absorber off-gas is 0.05%. In addition, it is desirable not to exceed the limit more than about 2% of the time.

Figure 5–5 shows the percentage of HCN in the absorber off-gas for a typical 24 hours of operation. The total standard deviation is 0.0044%. The best standard deviation with ideal feedback control (Equation 2–25) is not significantly better. The average percentage of HCN is 0.032%, which is well below the limit. It indicates that the water recirculation rate can be reduced to save steam without violating the specifications for HCN in the off-gas. Equation 2–14 and Table 2–1 estimate how often the HCN limit is exceeded.

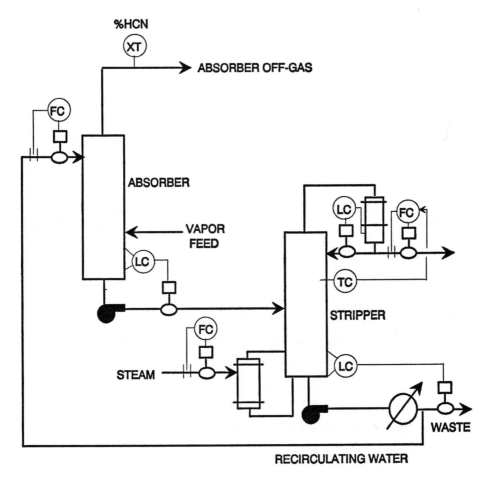

Figure 5-4   An Absorber/Stripper System

The fraction of HCN that exceeds the high limit

$$= 1 - \Phi(Z)$$

$$= 1 - \Phi\left(\frac{X_H - \bar{X}}{S_{\text{tot}}}\right)$$

$$= 1 - \Phi\left(\frac{.05 - .032}{.0044}\right)$$

$$= 1 - \Phi(4) < 0.0002 \text{ or } < 0.02\%.$$

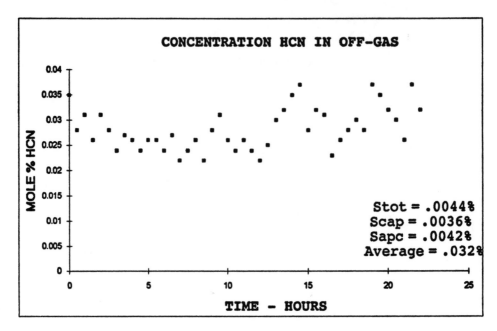

Figure 5–5   An Example of HCN Concentration Control

Clearly, water flow is too high, and it can be reduced until the percentage of HCN exceeds the 0.05% limit only 2% of the time.

The fraction of HCN that exceeds the high limit

$$= 1 - \Phi(Z)$$

$$= 1 - \Phi\left(\frac{X_H - \overline{X}}{S_{tot}}\right) = .02$$

Thus,

$$\Phi(Z_H) = 1 - .02 = 0.98.$$

From Table 2–1,

$$Z_H = 2 \text{ at } \Phi(Z) = 0.98$$

Solving for $\overline{X}$:

$$Z_H = \frac{X_H - \overline{X}}{S_{tot}} = 2$$

So

$$\overline{X} = X_H - 2\,S_{tot} = .05 - 2(.0044) = 0.041\%$$

By allowing the percentage of HCN to exceed the limit 2% of the time, the operating target for the percentage of HCN can be moved from 0.032% to 0.041%. The energy savings are calculated from the following operating data.

Recirculating Flow Rate	426M lb/hr
Stripper Steam Flow Rate	48M lb/hr
% HCN in Off-Gas	0.032%
Steam Variable Cost	$ 2.30/1000 lb
Operating Time	8000 hr/yr

$$\text{Energy savings} = 0.01\% \text{ increase in } \%HCN \times \frac{426M \text{ lb/hr water}}{0.032\% \text{ HCN}}$$

$$\times \frac{48M \text{ lb/hr steam}}{426M \text{ lb/hr water}} \times \frac{\$2.30}{M \text{ lb/hr steam}}$$

$$\times \frac{8000 \text{ hr}}{yr}$$

$$= \$276,000/yr$$

Prioritizing Improvements for Reduced Resources

The mid-1980s began a period when companies substantially reduced personnel to bring fixed costs more in line with global competition. As a consequence, plants lost many experienced and competent engineers, operators, technicians, and mechanics who had been responsible for operating and supporting the process. Many process improvement projects had to be put off because of the lack of experienced personnel to carry them out, even though capital was available.

In this environment, people have to learn to work smarter, not only in the intellectual sense, but in doing the things that will produce the most benefits. This chapter describes a methodology that identifies which opportunities will gain the biggest payback for the manpower expended. It assumes that manpower availability is an overriding constraint in getting process control projects completed.

OPPORTUNITY SELECTION

The methodology makes a high-spot estimate of the expected benefits for the control improvements and estimates the resource requirements in terms of the cost of specialists to develop, design, implement, and commission the improvements. The ratio of the potential benefits, or stake, to the resources is called the project selection index (PSI).

$$PSI = \frac{Stake}{Resources} \qquad (6-1)$$

The PSI measures the *return on resources*. It is a way to screen or prioritize projects or technical programs to gain the greatest return on scarce resources. Capital expenses are not included in the index, but may be estimated separately as another indicator of the feasibility of the improvements.

The methodology has been applied to over 100 processes. The average PSI was about 3, although the range was from less than 1 to over 20. This says that on average, process control specialists should be expected to achieve $300,000/yr in benefits through process control improvements for each man-year of effort.

PLANT SURVEY

The PSI methodology is implemented by on-site meetings or surveys involving people who are familiar with the process being surveyed (e.g., process engineers, operators, and mechanics). A process control specialist who may be associated with the process or from an engineering group is also needed to estimate the kinds of control improvements and resources required. The survey can be conducted over a 2-to-3-day period if only high-spot and rapid estimates are needed. These are useful to prioritize technical programs. If better estimates are needed to support capital projects, a more thorough study may be warranted.

The team should have a leader who is responsible for organizing the survey and collecting data for the team to review. Depending on familiarity with the process, Table 6–1 lists the kinds of background information that may be useful. The information should be distributed well ahead of the meeting so the team can prepare.

Table 6–2 shows the steps involved in the survey. The process is broken down into discrete units or operations, all of which impact costs and represent potential savings. The team leader or process engineer reviews the key business and manufacturing goals, and leads a discussion on the manufacturing operation. The costs associated with poor control are discussed, and the team estimates the savings potential of improving the measurements and control strategies. The methods presented in the previous chapters can be used to develop the estimates ahead of the survey, or it may be preferred to make on-the-spot estimates based on the team's operating experience.

TABLE 6–1   PSI SURVEY BACKGROUND INFORMATION

- A general description of the process: chemistry, unit operations, and key business drivers
- A simplified flow sheet or process schematic showing important flow rates and process conditions
- A control diagram showing the major control loops in the existing control schemes and noting any that are considered inadequate or troublesome
- Cost sheet data showing the unit values of raw materials, energy, and intermediates
- Performance data showing yield losses, utility losses, and other cost penalties
- A list of ongoing technical programs, goals, and suggestions for process control improvements

The process control specialist(s) conceptualize the kinds of improvements that will be needed for each opportunity. Improved controls may consist of simply tuning up the existing controls, in which case manpower is minimal. More resources will be needed if new strategies or improved process measurements are needed. If too little is known about the process dynamics (how the process responds over time to upsets), a mathematical model or simulation may be useful to gain the process understanding necessary to intelligently design an improvement.

If new control hardware and software are needed, resources will be used to provide detailed design. In some cases, it would be wise to include training program development and allowances for followup work to make sure the benefits will be sustained.

ESTIMATING THE STAKE

The economic benefits or stake for improved control is made up of several components.

$$\text{STAKE} = Y + E + P + Q + W \tag{6–2}$$

where

Y = annual savings from yield improvements
E = annual energy cost reduction

TABLE 6–2   PSI SURVEY STEPS

1. Draw boundaries around the process to include all units that affect operating costs and the business goals, and where control improvements can be made to gain benefits.
2. Identify groups of equipment that can be treated as individual processing units or process steps.
3. Analyze operating units to predict the impact improved control will have on:
   - Reducing yield losses
   - Reducing energy consumption
   - Improving product quality
   - Increasing throughput
   - Reducing waste and emissions
   - Improving cycle time
   - Improving uptime
4. Estimate the stake for each unit.
5. Estimate resource requirements for each improvement.
   - Design/implement/commission control strategies
   - Develop/design/install process analyzers
   - Define/install new control hardware and software
   - Develop/conduct training for operators, engineers, and mechanics
6. Calculate the PSI for each unit to highlight the most profitable opportunities.

P = annual value of increased production
Q = annual value of improved quality
W = annual savings from reduced waste treatment

Some of the equations have already been presented in Chapter 5 and are repeated here for convenience.

1. Annual Savings from Yield Improvements

The reduction in physical yield losses can be calculated by

$$Y1 = \left(\text{Flow rate} \frac{\text{units}}{\text{hr}} \right) \times \left(\frac{\% \text{ component lost}}{100\%} \right) \times \left(\frac{\% \text{ reduction}}{100\%} \right)$$
$$\times \left(\frac{\$}{\text{unit}} \right) \times \left(\frac{\text{hr}}{\text{yr}} \right)$$

The reduction in raw material loss due to chemical yield can be calculated by

$$Y2 = \left(\text{Raw material flow rate} \frac{\text{units}}{\text{hr}} \right)$$
$$\times \left(\frac{\text{\% recoverable yield loss}}{100\%} \right) \times \left(\frac{\text{\% reduction}}{100\%} \right) \times \left(\frac{\$}{\text{unit}} \right) \times \left(\frac{\text{hr}}{\text{yr}} \right) \tag{6-4}$$

2. Annual Energy Savings

$$\text{Energy} = \left(\text{Energy usage rate} \frac{\text{units}}{\text{hr}} \right) \times \left(\frac{\text{\% reduction}}{100\%} \right)$$
$$\times \left(\frac{\text{variable cost}}{\text{unit}} \right) \times \left(\frac{\text{hours of production}}{\text{yr}} \right) \tag{6-5}$$

3. Annual Value of Increased Production

$$P = \text{Current annual production rate}$$
$$\times \left(\frac{\text{\% estimated increase in throughput}}{100\%} \right)$$
$$\times \left(\frac{\text{\% of time sold out}}{100\%} \right) \times \left(\frac{\$}{\text{unit of increased production}} \right) \tag{6-6}$$

4. Annual Value of Improved Quality

$$Q = \text{Annual production rate}$$
$$\times \left(\frac{\text{\% increase in earnings for higher quality}}{100\%} \right)$$
$$\times \left(\frac{\$ \text{ earnings}}{\text{unit of product}} \right) \tag{6-7}$$

5. Annual Savings from Reduced Waste Treatment

$$W = \text{Annual waste treatment cost} \times \left(\frac{\text{\% reduction in cost}}{100\%} \right) \tag{6-8}$$

ESTIMATING THE RESOURCES

The cost of process control resources can be estimated from Equation 6–9.

$$\text{RESOURCES} = Ec + Ea + Es \tag{6-9}$$

where

Ec = Cost to understand the process, design the required control strategies, and assist in implementation

Ea = Cost to develop and install analyzers and sensors

Es = Cost to specify and install new control hardware and software

1. Process Understanding and Control Strategy Effort

Ec accounts for the time it takes to design the control strategies and assist in implementation and commissioning. It also accounts for gaining process understanding through modeling.

$$Ec = (NL \times MYL \times \$/\text{man-yr}) \times (1 + Fp + Fc) \qquad (6\text{--}10)$$

where

NL = Number of key loops
MYL = Man-years per loop
Fp = Weighting factor to account for current process understanding
Fc = Weighting factor to account for available control know-how

a. Number of Key Loops

The number of key loops is a way to gauge the magnitude of the control improvement effort. These are the major loops that will be affected by the improved control strategy. For example, the key loops on a distillation column that would be affected by improving composition control are the top and bottom composition controls. For a reactor, the key loops might be reactor temperature and feed flow control.

b. Man-Years per Loop

This number indicates the time expended to design and implement a control strategy. The magnitude is a function of complexity and the type of hardware and software used for implementation. The guidelines in Table 6–3 should be used to define this parameter.

c. Manpower Cost (Dollars per Man-Year)

This is the fixed cost of salaries and benefits. A typical number is $100,000/yr.

TABLE 6–3 MAN-YEARS PER LOOP GUIDELINES

Application	Man-Years/Loop
Simple regulatory feedback control	0.05
Regulatory loop containing feedforward, ratio, and cascade functions	0.10
Complex loop containing overrides, calculated variables, automated startup or shutdown logic, batch logic	0.20
Advanced model-based controls	0.30

d. Level of Process Understanding—Fp

Fp reflects the current level of understanding of the process and how it responds dynamically to upsets. It also includes the availability of technical information about the process. If the process is not well understood, knowledge will have to be gained by modeling and simulations, plant tests, or other means. The values placed on *Fp* are shown in Table 6–4.

In many cases, plants have already made control improvements if the process is well understood, so additional improvements will require some process analysis (i.e., $Fp = 1$).

e. Availability of Control Know-How—Fc

If the required control improvement technology is already known, very little is required to gain similar benefits in another process beyond

TABLE 6–4 FP VALUES

Level of Process Understanding	Fp
Process well understood with all data and models developed	0
Most key steps understood, some need to be modelled	1
Some key steps understood, most need to be modelled	3
Process is a black box that needs to be completely characterized, and data and models developed	5

straightforward implementation. If this is the first time such a process has ever been automated to such a degree, the effort required is much more substantial (see Table 6–5).

2. Analyzer and Sensor Effort

If there is one area of control technology that is most often the root cause of poor control performance, it is process measurements. Although recent measuring devices have made great strides in performance, the application or maintenance of the device may be the problem. This parameter looks at the effort associated with measurements.

Ea is the cost of manpower to develop and install analyzers and sensors. If new technology must be developed, considerable effort will be required. An alternative to hardware is to develop an on-line model that estimates the variables needed for control. Although this approach usually requires less capital, it does require considerable engineering effort, and often specialized expertise. This on-line modeling effort is included in Ea rather than Ec. Ea can be estimated from Equation 6–11.

$$Ea = \text{(Cost to develop analyzers or sensors)}$$
$$+ \text{(Cost to install analyzers and sensors)} \qquad (6\text{–}11)$$
$$+ \text{(Cost to develop and implement on-line models)}$$

The guidelines in Table 6–6 estimate the cost to develop and install analyzers.

The cost to develop and install on-line models depends on how complex

TABLE 6–5   FC VALUES

Control Know-How	Fc
Strategies proven on a known similar process	0
Control strategies must be developed but control is straightforward	0.5
Control strategies must be developed and advanced control techniques are required	1
New advanced control techniques must be developed	2

TABLE 6–6 ANALYZER COSTS

Type	Develop Cost	Install Cost
pH, conductivity	$15M	$20M
GC, IR	$30M	$45M

they are. A model to predict a process variable such as a composition may be in the range of $25,000 to $50,000. Great progress is being made in applying some of the new modeling techniques, such as neural nets, partial least squares, and principal component analysis, to chemical engineering problems. These models are promising for the more difficult nonlinear processes, which are almost impossible to model with traditional first principles methods.

3. Control System Effort

Es is the effort to design and install new control hardware if required. On a per-loop basis, about two man-weeks are required to replace or upgrade control hardware.

Example 6–1

The following example illustrates the PSI method for a Dowtherm™ heater. Hot Dowtherm from an oil-fired heater supplies heat to several distillation columns. The Dowtherm flow controllers to the reboilers are usually in manual mode because the oil-fired heater cannot handle the swings in flow that sometimes occur when the flow controllers are in automatic mode.

Opportunity for Improved Control

- Improve the oil-fired heater controls to allow automatic control of the Dowtherm flows to the columns.
- Implement fuel/air ratio control with excess oxygen trim and flow control of Dowtherm to each column to reduce fuel consumption.

STAKE: Reduce fuel costs by 20%.

$$
E = \left(\text{Annual production rate} \frac{\text{lb}}{\text{yr}} \right) \times \left(\frac{\text{fuel cost}}{\text{lb production}} \right)
$$
$$
\times \left(\frac{\% \text{ reduction in fuel}}{100\%} \right)
$$
$$
= 10\text{MM lb} / \text{yr} \times \$0.05 / \text{lb} \times 0.2 = \$100\text{M/yr}
$$

Resources:

- Number key loops = 4
- Man-years per loop = 0.1
- $ per man-year = $100M
- $Fp = 1$
- $Fc = 0$

$$Ec = [4 \times 0.1 \times \$100M] \times [1 + 1] = \$80M$$

$$Ea = \$25M$$

Total resources $= \$105M$

$$PSI = \frac{\$100M}{\$105M} \sim 1$$

A PSI = 1 is quite low. An average value is around 3.

7

Implementing Automatic Controls for Reducing Variability

Up to this point, the focus has been on process analysis to identify opportunities for and the stakes in improving process control. Little has been said about how to apply specific automatic control algorithms to achieve reduced variability. The topic of this chapter, therefore, is to assess the capability of various control algorithms and describe some of the problems that can hinder the ability to reduce variability. Before getting too deeply into the subject of automatic process control, some basic concepts are first introduced.

AUTOMATIC CONTROL STRATEGIES

To reduce variability, we should first attempt to remove the sources of variability if feasible. Why burden control with compensating for disturbances when they can be removed? This is the thrust of statistical process control. However, not all causes of variability can be eliminated at the source, so we attempt to reduce their effect by compensation.

There are some conditions necessary to have effective compensatory control. Of course, we need to be able to measure the effect of the disturbance on some process variable that we can then control. Also, we need to have a manipulative variable close enough to where the disturbance enters the process that it can counteract the disturbance effectively. The manipulative variable should also have a significant effect on the process variable we are trying to control. This is what control engineers call *process gain*. Given these conditions, we should be able to design a control strategy that compensates for the disturbance.

The primary control structure is a feedback control loop in which we measure the process variable, compare its value to the desired value (setpoint or aim), and then calculate a control action based on the amount of deviation. The feedback loop is called that because the measurement is fed back to the controller before any control action is calculated. One drawback to feedback control is that a deviation must occur before the control does any compensation. This chapter shows that feedback control only reduces variability up to a point because of that limitation.

Sometimes feedback controllers are arranged in series, or *cascade*, where the first or primary controller output becomes the setpoint to the secondary controller. This scheme has two controlled variables; for it to work properly, the secondary variable response must be faster than that of the primary variable. Cascade control is a useful way to remove variability that occurs in the secondary loop before it affects the primary variable.

The last basic strategy is called *feedforward control*. If the disturbance can be measured directly, a compensatory control action can be fed forward to adjust the manipulative variable before the disturbance is felt by the primary variable. Feedforward is very effective for reducing variability, and if done properly, it is much more effective than feedback control. The feedforward action has to be timed in such a way that it will exactly cancel the disturbance effect. If the feedforward action is too early or late, not only will it not cancel the effect of the disturbance, it may even introduce more variability. Therefore, feedforward control requires a model of the process where there are dynamic effects so that it can be correctly synchronized to the process response.

This discussion assumes that there is a measurable process variable. Sometimes an appropriate measurement does not exist. Then we try to infer the measurement by a mathematical model based on a few indirect measurements. These *predictive* models are the basis of model-based controllers. They are discussed in the following chapter.

MINIMUM VARIANCE CONTROL

Chapter 2 presented a simple metric, S_{apc}, for the minimum standard deviation obtainable with feedback control. S_{apc} can theoretically be reached using a feedback controller specifically designed to achieve minimum variance. The design of such controllers has been the subject of many papers. Harris (1982) discusses one kind of minimum variance controller that is based on statistics and designed by time series analysis techniques. This controller is a good example of how statistics and process control can be combined in a synergistic way to design high-performance controllers. The prerequisites for designing the controller are that the process model is known and the disturbance can be modeled as an autoregressive-integrated-moving average (ARIMA) time series. The disturbance is modeled as having a random, meandering motion very much like many kinds of disturbances we encounter in real processes. The process and disturbance models are directly used in the synthesis of the control algorithm.

Most controllers designed by classical process control techniques differ in that they assume disturbances are not stochastic but deterministic, that is, they are not random but have predictable forms such as steps or ramps. These model-based controllers are also synthesized from assumed forms of the disturbances and process dynamics.

Although there are important applications where high-performance model-based controllers are really needed to get satisfactory control, most situations we encounter in industry can still be handled well by the more conventional types of controllers, namely, PID controllers. These controllers are not synthesized from models but have a fixed structure. Since the vast majority of controllers in the field are of the PID type, an obvious question is: How close do these controllers come to achieving ideal performance in terms of minimum variability? Furthermore, since the minimum variability metric is based on feedback control, one would like to know how much more reduction in variability is possible with the addition of feedforward control. The purpose of this chapter is to explore these and several other implementation issues.

PID CONTROL ALGORITHM

PID (proportional-integral-derivative) controllers have been the mainstay of the chemical processing industries since the 1930s. The structure of the PID algorithm is optimal for many kinds of processes encountered in the processing industries. McMillan (1990) points out in his book on controller tuning that except for processes dominated by deadtime, PID controllers are really the best for counteracting unmeasured load disturbances. Deadtime refers to a situation in which the process response is delayed, and it is one of the most difficult situations to deal with in control. The fact that most control situations are not dominated by deadtime might explain why the PID has been so successful over the years.

The PID controller is made up of three separate modes of control: proportional, integral, and derivative. Each mode has its own unique action. The proportional mode adjusts the controller output based on the deviation *e* between the controller's setpoint and the feedback process variable; the greater the deviation, the greater the output change. The amount of control action is determined by the proportional gain or, in some controllers, the proportional band, which is related inversely to gain.

The proportional mode by itself has often been used to control levels in tanks. Suppose the controller manipulates the outlet valve. If the inflow increases, the level will rise, and the controller will increase the outflow until the outflow balances the inflow. When this occurs the level stops moving, and so does the controller. However, the level will have reached a new value that is different from the original setpoint.

The integral mode of the controller, also called reset, adjusts the output based on the integral of the deviation from setpoint. It integrates the deviation from setpoint until either the process variable returns to the setpoint or the controller output reaches its high or low limit. The speed of the integration depends on the integral time parameter whose inverse multiplies the integral term. Integral mode is always necessary if you want to maintain a specific setpoint for the process variable. Sometimes integral is used by itself, but usually it is used in combination with proportional (PI).

Using the level control example with a PI controller, the integral mode actually drives the outflow to momentarily change more than the inflow change in order to get the level back to setpoint. One precaution with PI level controls is that if you have a series of vessels all having level controllers, the inflow disturbance will be amplified by each succeeding vessel, which increases variability. The amount of amplification depends on how tightly the controller is tuned, that is, the value of the integral time parameter. For that reason, many level control applications use only the proportional mode. If the vessel is large enough, it usually is not that critical to maintain a certain level anyway. A technique called *averaging level control* adds only enough integral action to slowly bring the level back to setpoint. This way, the tank can act as a surge vessel to attenuate variability in the inflow so that the outflow can be smooth and gradual.

Finally, the derivative mode adjusts the output based on how fast the process variable is changing. The faster the variable changes, the more control action is applied. The amount of derivative action is determined by the derivative time parameter that multiplies the derivative term. The trick in achieving a good process response and stable control is in setting the tuning parameters properly. Since the derivative mode adds a third tuning parameter, it is often not used. Getting the proportional and integral set right is usually gratifying enough for the control engineer. However, the derivative mode does have value in speeding up the response in many situations, particularly in compensating for secondary lags; for example, the lag in a thermocouple installation.

The textbook expression for the discrete PID controller is given by Equation 7–1.

$$M_n = Kc \left[e_n + \frac{Ts}{Ti} \sum_{i=0}^{n} e_i + \frac{Td}{Ts} (e_n - e_{n-1}) \right] \qquad (7\text{--}1a)$$

where

M_n = controller output
e = $X_{sp} - X$, the setpoint minus the process variable
Ts = sampling period
Kc, Ti, and Td are tuning parameters.

We can also express the controller in terms of the previous controller output, M_{n-1}.

$$M_n = Kc \left[(e_n - e_{n-1}) + \frac{Ts}{Ti} e_n + \frac{Td}{Ts} (e_n - 2e_{n-1} + e_{n-2}) \right] + M_{n-1} \qquad (7\text{--}1b)$$

Actual implementation of the PID is somewhat more complex and varies from vendor to vendor, but Equation 7–1 expresses the main idea. For example, the derivative mode usually acts only on the process variable and not on the deviation to avoid a spike in the output when the setpoint is changed. In addition, the derivative is usually filtered to provide a smoother response. One other point is that the controller can be direct-acting or reverse-acting, depending on the direction in which the process responds to the manipulated variable. A direct-acting controller increases its output if the process variable increases, whereas a reverse-acting controller decreases its output if the process variable increases.

The responsiveness of the controller is determined by the tuning parameters *Kc, Ti,* and *Td.* The amount of proportional action is determined by the proportional gain, *Kc.* Note that *Kc* also affects the integral and derivative modes. The integral time, *Ti,* which has units of minutes per repeat, determines the amount of integral action. Finally, *Td,* which has units of time, is the derivative time, which determines how much derivative is applied.

Since the modes can be used in almost any combination, this method provides the opportunity to select the control structure that precisely matches the process control situation. For example, if the process resembles a single lag, that is, the process changes exponentially with no deadtime to a new steady-state value following a step input, the PI structure is optimal. For a lag with deadtime, where the lag dominates, the PID form is best. For a process that exhibits an immediate change to a new steady state with no dynamics, the I form is best. To reiterate, most applications use only PI because D can be tricky to tune, especially if there is a fair amount of measurement noise. PI is adequate for most applications, and D is added to some loops to get higher performance.

As you might expect, there are many techniques for tuning PID. Just about every process control book will have a section on controller tuning. McMillan (1990) is completely dedicated to tuning. One of the earliest and best known techniques is the Ziegler-Nichols (Z-N) tuning method [Ziegler 1942]. It sometimes results in tuning that is too aggressive for an actual application, but it is a good benchmark and often serves as a standard of comparison. Aggressive means that large control actions are called for, which could cause upsets in other control loops. However, because Z-N tuning is the standard, it is used in the examples that follow in assessing control performance. Later, a PID controller with Z-N tuning is compared to minimum variance control.

SAMPLING PERIOD

A parameter that affects the performance of a control loop is the time interval between making control adjustments (i.e., the sampling period). Modern digital control systems are designed to approach continuous control by having very short sampling periods of much less than one second. Most measurement devices for level, flow, pressure, and temperature are also basically continuous.

Processes typically react in a longer time frame, so the performance of the loop is generally not degraded by how fast the adjustments are made. Two exceptions are the anti-surge controls on a centrifugal compressor and temperature control on a continuous polymerization reactor. The response time for these processes may be in milliseconds. However, some analytical measurements are not continuous and therefore slow down the execution speed of the control loop. For example, a gas chromatograph analyzer that measures concentrations may operate on a cycle time of 10 to 15 minutes.

Another example where the control is delayed is when deadtime exists in the measurement installation. Deadtime refers to a time delay during which the measurement instrument is unable to detect a change in the process variable, even though a controller adjustment may have been made. Such is the case with a viscometer, which draws a sample of fluid from the process through a tube. The deadtime corresponds to the time it takes for a molecule to make its way down the tubing. A long length of tubing can add a significant amount of deadtime, which will degrade performance and ultimately result in greater process variability.

Both long sampling periods and deadtimes affect how much we can improve variability. A simulated process will be used to illustrate. Figure 7–1 shows a process for blending two ingredients that make up the feed stream to a reactor. From the point of blending, the fluid flows through a long pipeline, representing a process deadtime, to a mixing tank, which represents a process lag. The lag simulates the inertia or resistance to change in concentration. The process can be described by a first-order ordinary differential equation of the form

$$\tau \frac{dX(t)}{dt} + X(t) = Kp\, M(t - D) \tag{7-2}$$

X = process variable under control

M = manipulative variable

Kp = steady-state process gain

τ = process time constant in minutes

D = process deadtime, minutes

The response of X to a step change in M is described by Equation 7–3.

$$\Delta X(t) = Kp\, (1 - e^{-t/\tau})\, \Delta M(t - D) \tag{7-3}$$

X begins to change D minutes after M changes and reaches 63.2% of its final value τ minutes after that. The steady-state change in X to a unit-step change in M is Kp.

The concentration of ingredients is controlled at the tank exit by manipulating one of the flows. This is a so-called feedback control strategy in which the process variable, concentration, is measured and fed back to the concentration

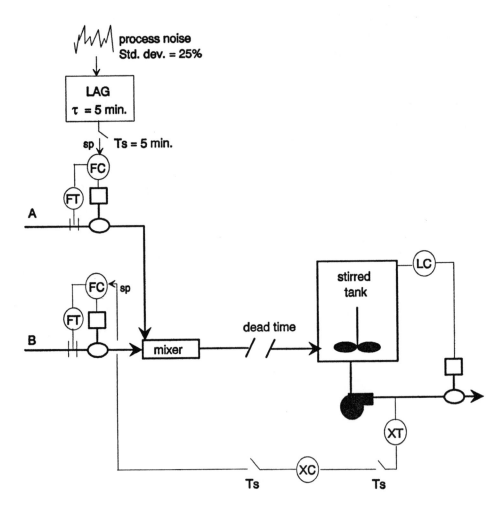

Figure 7–1   First-Order Process with Deadtime

controller (XC). The XC compares the variable to a setpoint and generates a control action. In this case, the concentration controller output is the external setpoint to a flow controller (FC). The FC adjusts the control valve until flow matches its setpoint. Since two controllers are in series, this is called a "cascade feedback control loop." The other flow is the source of disturbance. The upset is generated by a random, meandering setpoint on the flow controller. This example will show

how the achievable reduction in variability is affected by how frequently we can measure the concentration.

In the first case, a lag dominant process is considered. The level in the tank is set to achieve a process time constant (τ) of 10 minutes. The deadtime (D) is set by the length of pipe and flow rate to be one-tenth of the process time constant. This lag-dominant type of process characterizes many actual process responses in chemical plants. Some examples are the exit temperature in a heat exchanger, temperature on a distillation column tray, and reactor jacket temperature.

Figure 7–2 shows the open loop response in concentration as it is measured at several rates. The sampling period, *Ts,* is the time between taking measurements. There are large differences in how the data look. The short sampling peri-

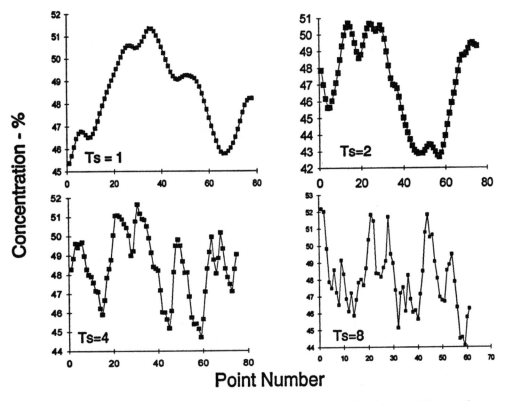

Figure 7–2 Open-Loop Response Data Showing Increased Randomness (Decreased Autocorrelation) as the Sampling Period Increases

ods capture the transient response very accurately, clearly showing the inertial effects of the process. Because of inertia, the data are not independent of one another but are affected by the previous data. This is what statisticians call autocorrelation in the data. Controllers are designed to overcome process inertia to speed up the response. Perfect control removes all the autocorrelation, and the resulting data are then random. This is the state of statistical process control.

For the longer sampling periods, the inertial effects are not as evident, and the data appear more and more random. If the process reaches steady state before the next sample is taken, the variations in the data are truly random because all the inertial effects have played out. For control to be effective, the data must be autocorrelated, that is, there must be inertia in the process.

As the data become random, the capability standard deviation coincides with the total standard deviation, and feedback control cannot reduce variability any further as long as it operates at the same frequency. This is illustrated in Figure 7–3, which is a plot of standard deviation versus the process time constant divided by sampling period (τ/Ts). The curves were generated for the open loop situation (i.e., without control). For short sampling periods (large values of τ/Ts) there is a large difference between S_{cap} and S_{tot}, and consequently between S_{apc} and S_{tot}, signifying a large opportunity to reduce variability by control. However,

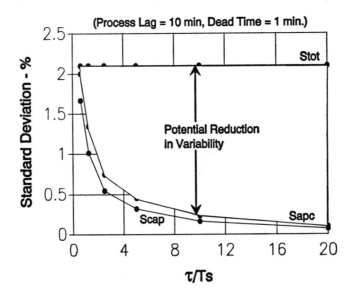

Figure 7–3 Potential Reduction in Variability as a Function of the Sampling Period

as the sampling period gets larger, the data become more random, and there is less opportunity for reducing variability. Finally, S_{cap} and S_{apc} approach S_{tot}, and there is no opportunity at all to reduce variability. Thus, in order to improve control, the data must be taken at a high enough frequency that inertial effects are present.

Figure 7–3 illustrates another point. As Ts exceeds $\tau/2$, you rapidly lose potential to reduce variability. Conversely, as Ts becomes less than about $\tau/4$, you do not have much to gain by reducing Ts further. To be safe, many practitioners sample 5 to 10 times per time constant.

This result brings to mind Shannon's sampling theorem, which has to do with how fast a signal needs to be sampled to adequately represent it. It says that to be able to completely recover the continuous signal from its sampled counterpart, the sampling frequency must be at least twice the highest frequency of the signal [Smith 1972]. In terms of the time constant, the sampling period must be less than half the smallest time constant of interest. Shinskey (1979) recommends that for the derivative mode of a PID controller to be effective, the sampling rate should be even faster: 60 to 125 samples per time constant.

The point of this discussion is that as Ts increases, the ability of feedback control to reduce variability is curtailed. This is another way of saying that feedback control effectiveness is limited by the sampling frequency. If an analysis of the process data shows that the minimum achievable standard deviation through feedback control, S_{apc}, is not good enough, one way to solve the problem is to install a faster-responding measurement instrument or have an on-line continuous model to predict the variable.

PERFORMANCE OF PID CONTROL

The concentration control loop is closed in Figure 7–1 to test how closely conventional feedback controllers can achieve the minimum variance predicted in Figure 7–3. PI and PID controllers, both with Ziegler-Nichols tuning, are compared. The tuning was calculated from the ultimate period and gain that were determined by Astrom's ATV method [Luyben 1990]. The ATV method generates a small cycle in the process variable, which corresponds to the ultimate period (Pu) and ultimate gain (Ku) of the process. The Z-N tuning is calculated from Table 7–1 for a lag-dominant process. (Note that these parameters will be much too aggressive for a process dominated by deadtime.)

The resulting tuning settings are shown in Table 7–2.

Figure 7–4 shows the standard deviations achieved with feedback control at

TABLE 7–1   Z-N TUNING FOR A LAG-DOMINANT PROCESS

Parameter	P	PI	PID
Kc	$Ku/2$	$Ku/2.2$	$Ku/1.7$
Ti		$Pu/1.2$	$Pu/2$
Td			$Pu/8$

the different sampling periods, Ts, for the lag-dominant process. The PI and PID controllers were synchronized to take control action every time they received a new measurement. The output was held constant between samples. This is referred to as *sampled data control*. Note that PID control with Z-N tuning came close to achieving minimum variance control. The PI performance was less, but would probably still be considered satisfactory in many situations.

Figure 7–4 cheats a little in that it shows that as τ/Ts decreases (Ts is large), the curves converge smoothly to the open loop total standard deviation curve. In actuality, as τ/Ts gets small, the controlled response exhibits more randomness. Feedback control becomes less effective because it is trying to control a random process. Table 7–3 gives the actual standard deviations.

At Ts = 16 minutes, the process dynamics or inertial effects have largely played out between samples, so the resulting standard deviations from control are random. The lesson here is that if you apply control to random data, control will be ineffective and perhaps even detrimental, especially if you try to get tight control. The percentage of reduction in standard deviation achieved by the PI and PID controllers are shown in Table 7–4.

The PID structure appears to be ideal for this type of process. Achieving the minimum variance depends on the controller tuning. Ziegler-Nichols tuning came close to minimum variance. McMillan (1990) has a table of tuning settings to

TABLE 7–2   Z-N TUNING FOR FIRST-ORDER LAG PLUS DEADTIME PROCESS (τ = 10 MINUTES, D = MINUTE)

			PI		PID		
Ts	Ku	Pu	Kc	Ti	Kc	Ti	Td
2	2.5	11	1.1	9	1.5	5	1.3
4	2.1	12	1	10	1.2	6	1.5
8	1.3	15	0.6	12.5	0.8	7.5	1.9
16	0.7	34	0.3	25	0.4	15	4

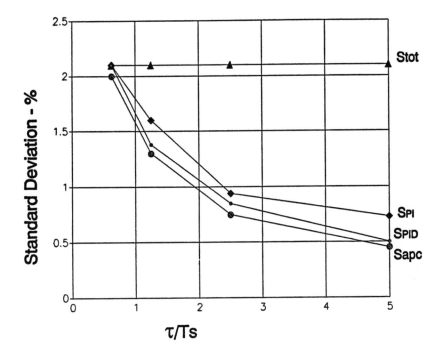

Process Lag = 10 min., Dead Time = 1 min.)

Figure 7–4   Performance of Sampled-Data PI/PID Controllers as a Function of the Sampling Period

achieve what amounts to minimum variance control. The tuning is based on minimizing the integral of squared errors, which is equivalent to variance. These settings result in very tight control, which may be necessary for quality control loops. The downside is that the manipulative variable has to be changed drastically, which may or may not be desirable. It depends largely on how other control loops in the process are affected, because we do not want to cause severe interactions.

This example showed that feedback control degrades as the sampling period increases. If it turns out that the potential reduction in variability as indicated by the difference between S_{tot} and S_{apc} is not sufficient for the control performance, there are a couple of alternatives. The first is to seek a new measurement device

TABLE 7–3 STANDARD DEVIATION FOR CONTROL OF A LAG-DOMINANT PROCESS

Ts	τ/Ts	S_{tot} (Open Loop)	S_{apc}	S_{PI}	S_{PID}
2	5	2.1	0.45	0.73	0.45
4	2.5	2.1	0.75	0.94	0.88
8	1.25	2.1	1.3	1.6	1.4
16	0.63	2.1	2	2.2	2.4

that has a short sample period. If that is not feasible, developing an on-line predictive model or inferential measurement that will operate at a much higher frequency should be considered.

DEADTIME-DOMINANT PROCESS

Another situation that limits the performance of conventional PI or PID controllers is a process in which the deadtime is greater than the process lag. Deadtime is the period of time between when the manipulative variable is adjusted and the measured variable begins to change. Deadtime is the mortal enemy of control, and process designs should minimize or eliminate deadtime whenever feasible. Deadtime requires that the controller tuning be slowed down so that it does not over react and cause the controls to go unstable. Note also that the derivative mode becomes useless in deadtime processes. Thus, performance is degraded.

It is not a good practice to try to compensate for deadtime by control if it can be avoided. One recommended practice is to locate the manipulative variable as close as practical to the controlled variable.

TABLE 7–4 PERCENTAGE OF REDUCTION IN STANDARD DEVIATION

Ts	τ/Ts	PI%	PID%
2	5	66	78
4	2.5	55	58
8	1.25	23	34
16	0.63	0	0

In general, control should not be required to compensate for impossibly complex equipment designs if the process can feasibly be simplified. Even if the initial cost is greater for the simplified process because more vessels are required, it may result in lower operating costs later on, which might compensate for the initial increased investment. One example comes to mind. Separating three components in a liquid stream typically requires two distillation columns, although it may be possible to accomplish the separation in one column that has a top, bottom, and sidestream product. Control on the latter column will be difficult if high purities are required. In this case, a model-based multivariable strategy may be necessary. The column should be simulated to make sure it is controllable in the face of expected upsets.

Figure 7–5 shows the potential reduction in variability achievable with better feedback control for the process in Figure 7–1 for various deadtimes. Just as

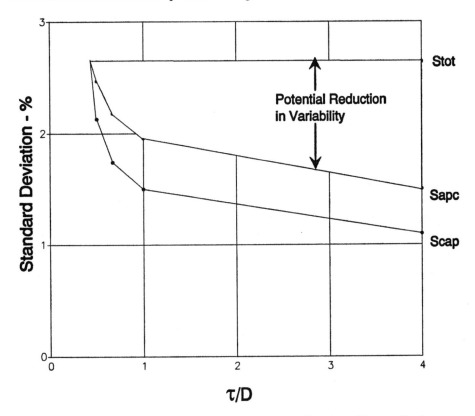

Figure 7–5   Potential Reduction in Variability as a Function of Process Deadtime

with sampling periods, increasing deadtime reduces the potential reduction in variability through better feedback control.

Figure 7–6 illustrates the results of applying a PI controller to the deadtime-dominant process. The tuning used is recommended by McMillan (1990) for a deadtime process. The PI works reasonably well for a moderate deadtime ($\tau/D >$ 1), but as deadtime surpasses the time constant of the process, PI performance degrades.

One of the classic ways in which deadtime is dealt with is deadtime compensation. A well-known algorithm for doing this is called the Smith Predictor [Smith 1959], which is shown in Figure 7–7. The Smith Predictor is used in conjunction with a PI controller. The predictor removes the deadtime from the feedback measurement, so the controller can be tuned as if deadtime were not present in the process. However, the deadtime will still be manifested in the actual process response.

The Smith Predictor works as follows. The controller output is input to a process model that excludes deadtime. The model output that is time-delayed is subtracted from undelayed model output, and the result is then summed with the

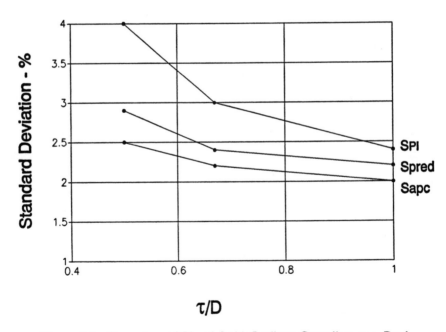

Figure 7–6 Comparison of PI and Smith Predictor Controllers on a Dead-timeDominant Process

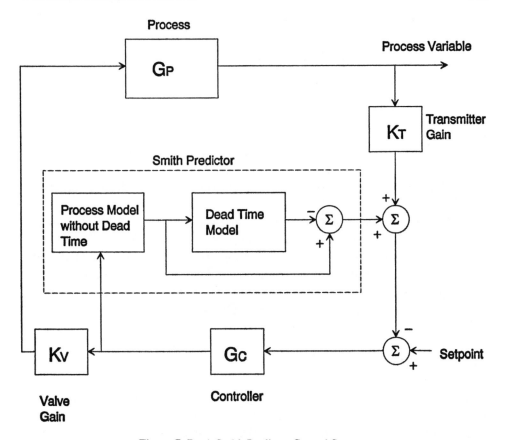

Figure 7–7   A Smith Predictor Control Strategy

feedback measurement. The effect is to artificially cause the feedback to respond immediately, just as if the deadtime were not present in the process. When the actual process response begins to show up, the model response decays away. As long as the model is accurate, this technique works pretty well. Actually, this technique is not confined to compensating for deadtime. By changing the model, it can be used to compensate for other types of undesirable behavior like the inverse response in boiler drum level control [Shunta 1984].

Figure 7–6 also shows the performance S_{pred} of the Smith Predictor using tuning that McMillan (1990) recommends for a deadtime process. The Smith Predictor gains an improvement in variability, especially as deadtime gets larger ($\tau/D < 0.8$). Table 7–5 shows the percentage of improvement of the Smith Predictor over straight PI control. This demonstrates that when deadtime dominates,

**TABLE 7–5   IMPROVEMENT OF PI
CONTROL WITH SMITH PREDICTOR**

τ/D	Standard Deviation		Improvement (%)
	PI	Smith Predictor	
1.00	4	2.9	28
0.67	3	2.4	20
0.5	2.4	2.2	8

conventional PI controllers are not adequate, and some sort of deadtime compensation controller is necessary.

Many papers have been written over the years evaluating the performance and robustness of the Smith Predictor [Meyer 1976, Schleck 1978]. The scope of this book does not include getting into that kind of detail, so the interested reader is encouraged to read the papers. Suffice it to say that the Smith Predictor is a pseudo-standard for deadtime compensation, but it is not bulletproof and should be applied with caution and with an understanding of its shortcomings.

FEEDFORWARD CONTROL

Feedback control has the disadvantage that it has to sense an error before it can react. If the disturbance can be measured, feedforward control can be applied. It calculates a control action directly from the disturbance itself without waiting for its effect on the process variable. Thus, it adjusts the manipulative variable directly to counteract the disturbance effects. In the process of Figure 7–1, feedforward control is easy; you just ratio the manipulated flow to the disturbance flow (Figure 7–8). This keeps concentration essentially constant, and variability is reduced to zero. Note that feedforward control is highly effective in processes with deadtime.

In other cases, feedforward may not be so straightforward. There may be some dynamic response differences in how rapidly the disturbance and manipulative variables affect the controlled variable. The feedforward controller must then account for these differences, or else the control action will occur too soon or too late, and variability will be worse than without feedforward control. The design of a feedforward controller that takes dynamic effects into account is shown in Figure 7–9. The process model Gp, which relates the process variable and the manipulative variable, and the model GD, which relates the process variable and the disturbance, must be known. The feedforward dynamic compensator GFF, which keeps the process variable constant, is just the ratio of GD/KTD KV GP.

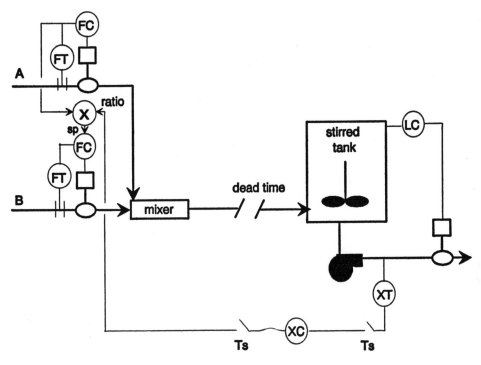

Figure 7–8 Feedforward Ratio Control for Controlling Composition

KTD and *KV* are the disturbance transmitter gain and valve gain, respectively. The output of the compensator is combined with the feedback controller *Gc* output. In this case, the controller outputs are summed. In the ratio control case, the feedforward output is multiplied by a parameter whose value is the ratio of the manipulated flow to the disturbance flow and becomes the manipulated flow setpoint.

An example of the summing feedforward strategy is shown in Figure 7–10 for a boiler drum level control. Since boiler feedwater is added to meet the steam demand, steam flow sets the feedwater controller setpoint. However, transmitters are not always accurate and calibrated properly, so the feedback level controller biases the feedforward signal to make sure the material balance is maintained.

A third type of feedforward control is called impulse feedforward. This is a feedforward signal that makes a step change and then decays away, just like the derivative mode. It has been used whenever a small momentary kick is required in the signal. An example is shown in Figure 7–11. A vapor stream is ratioed to the other feed to keep the mass flows in proportion. The pressure in the vaporizer will drop when the vapor flow increases. The feedforward action anticipates the

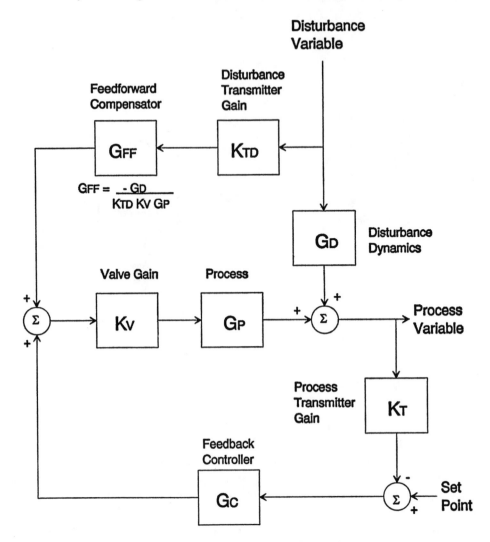

Figure 7–9   A Generic Feedforward Control Strategy

reduction in pressure and opens the steam valve immediately to keep pressure as constant as possible. In the long run, we want the slower pressure controller to have command so that the feedforward signal decays away exponentially as the pressure controller picks up the load. Most control systems have an impulse function that can be configured for this purpose.

In many kinds of operations, we routinely make good use of feedforward control. Some popular examples are boiler drum level controls, distillation con-

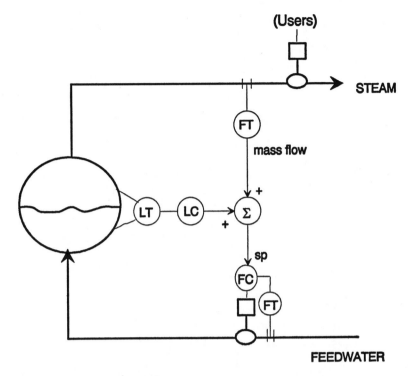

Figure 7-10   Feedforward Level Control for a Boiler Drum

trols, and reactor feed flow controls. Often, however, the dynamics for feedback control alone are fast enough that feedforward does not improve control enough to justify the additional complexity. Since feedforward control depends on having a good model of the effect of the disturbance and manipulative variables on the process variable, nonlinearities in the process that make a model inaccurate in some operating conditions may degrade performance. However, feedforward control should always be considered as an option, because if it is done correctly it can greatly reduce and even eliminate variability due to the disturbance.

CASCADE CONTROL

Cascade control is another mechanism for reducing variability and getting a faster speed of response. It is illustrated in Figure 7-1 by the XC (primary) that is cascaded to the FC (secondary). A cascade arrangement will reduce variability in

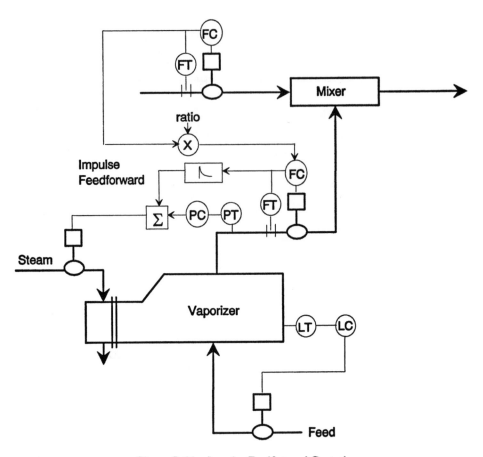

Figure 7–11   Impulse Feedforward Control

the primary variable caused by fluctuations in the secondary variable through the compensating action of the secondary controller. Cascade loops can be used if the response time of the secondary variable is much faster than the primary variable. For the loop to be stable, the secondary variable response should be at least three, and preferably five to ten, times faster than the primary controller. As a footnote, it is important that operators understand this concept well. Many times we have found that cascade loops did not function simply because the secondary controller was tuned more slowly than the primary controller.

Cascade control strategies have two advantages: they remove variability in the secondary variable, so the primary variable has less variability; and they speed up the response of the loop. Liptak (1986) reports that the period of oscillation is usually reduced by as much as 50% when using a cascade arrangement. That, along with the tighter tuning allowed, represents a fourfold improvement in the speed of response.

The example of temperature control in a distillation column demonstrates the first advantage. Figure 7–12 shows a binary distillation column with a top and bottom product and liquid feed. The main product is the bottom product, so its

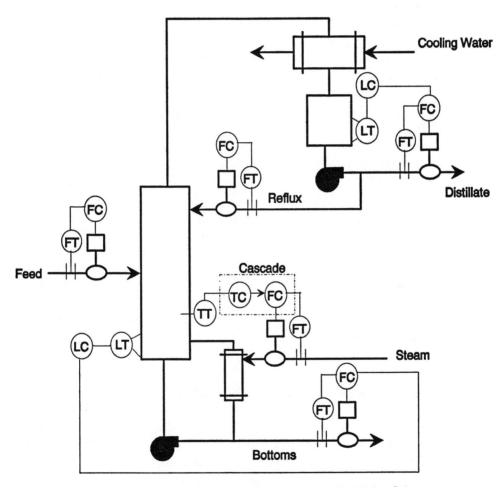

Figure 7–12   Temperature Cascade Control on a Distillation Column

composition is controlled indirectly by maintaining a specified temperature on the second tray from the bottom. The temperature is controlled by varying the amount of boilup generated in the reboiler by adjusting steam flow. In this case, the steam header has a random variation in pressure that upsets the steam flow and causes variability in the temperature. Three cases are considered:

- Case 1—open loop (temperature controller in manual)
- Case 2—temperature controller adjusts steam flow valve directly
- Case 3—temperature controller cascaded to steam flow controller

The response of temperature and steam flow is shown for the three cases in Figure 7–13. In Case 1, the temperature has a meandering behavior because of the variation in steam flow. In Case 2, the temperature is now under control, but still has a slight variation of about half a degree. Steam flow still has about the same amount of variability, although it does not have long-term drift as in Case 1. Case 3 represents very good control. The only variability is process noise. Ziegler-Nichols tuning gets pretty close to minimum variance.

Figure 7–14 quantifies how variability is reduced. In Case 1, the total standard deviation of temperature S_{tot} for the open loop situation is around 0.8 degrees. The minimum standard deviation for feedback control S_{apc} is estimated to be about 0.04, which represents a very large opportunity to reduce variability. By going to straight feedback control (Case 2), standard deviation was reduced to 0.12, which is close to the minimum even with Ziegler-Nichols tuning. However, going to cascade control (Case 3), the standard deviation was reduced below the minimum for straight feedback control. A little better than the minimum was achieved because the minimum standard deviation was calculated from data that was recorded every minute, but the controls were executed on a five-second cycle. This makes the point, again, that the sampling period plays a part in what you can achieve in reduced variability. Even so, it clearly illustrates the advantage of cascade control in minimizing variability in the secondary variable.

USING PROCESS VESSELS TO REDUCE VARIABILITY

A tried-and-true remedy for reducing variations in product quality has been to use tankage for blending good product with off-spec product to improve product consistency. More and more, however, process engineers attempt to reduce investment in capital equipment by eliminating unnecessary tankage, which also minimizes cash tied up in inventories. Ideally, by applying the concepts in this

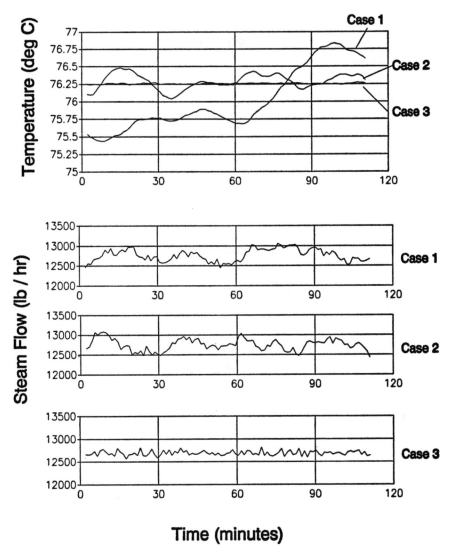

Figure 7–13 Comparison of Open-Loop, Single-Loop, and Cascade Temperature Controls on a Distillation Column

book, the process will be controlled so well that product will meet specifications without having to resort to blending. Therefore, addressing the use of blending tanks to reduce variability seems counterproductive. On the other hand, it is legitimate to use the surge capacity of process vessels to dampen upsets. This is done

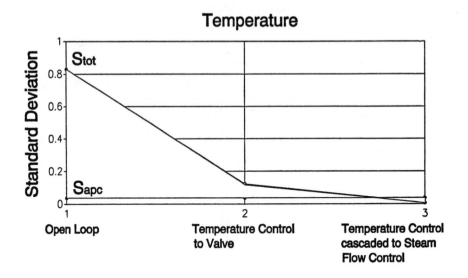

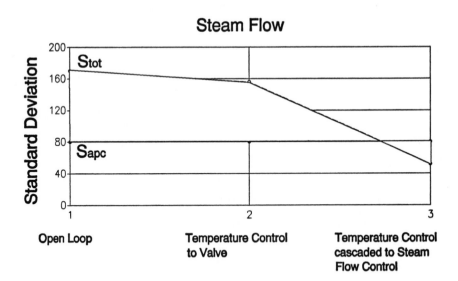

Figure 7–14   Variability Improvements with Single-Loop and Cascade Temperature Controls on a Distillation Column

by allowing levels to drift within the bounds of high- and low-level alarms so that flows between vessels are smoothed.

If a PI level controller is tuned tightly to hold a constant level, the manipulated flow will amplify the upset coming into the vessel. In other words, the controller must change the manipulated flow to a greater extent than the disturbance flow in order to bring the level back to setpoint. The amplification becomes greater as the level controller is tuned for a faster response. If several vessels are arranged in series, the amplification grows with each succeeding vessel, and instability can occur at some point. What we want to do is tune the controllers as loosely as possible so that the flow upsets are attenuated. This type of level control is called *averaging level control* because the level is maintained at setpoint on average, but at any point in time the level may drift away from setpoint to take up the surge in flow [Buckley 1985].

The previous distillation column example is used to illustrate the advantage of averaging level control. A feed tank is added, and the feed to the column is manipulated to control the level in the tank. The effect on tray 2 temperature control from flow changes into the tank and level control tuning is shown. Two sets of tuning are used (Table 7–6).

The feed tank is subjected to a 20% increase in inflow. Figure 7–15 shows the time response with both sets of tuning parameters. Note that the tight tuning results in a relatively large overshoot in feed flow, but the level only deviates a few percent. The column tray temperature deviates about a degree. With the averaging tuning, the feed overshoot is much less, and the level deviates over 10%. However, the column temperature controller is able to do a much better job of controlling because the feed upset is attenuated. The specifics are shown in Table 7–7.

This example clearly shows the benefits in using the surge capacity in vessels to attenuate flow upsets and make it easier for product quality control loops, such as the distillation column temperature, to maintain control.

TABLE 7–6   COMPARISON OF LEVEL TUNING SETTINGS

	Tight Level Tuning	Averaging Level Tuning
Kc	1	0.4
Ti	1 min/repeat	30 min/repeat
Closed-Loop Time Constant	6 min	15 min
Damping Ratio	0.3	1.0

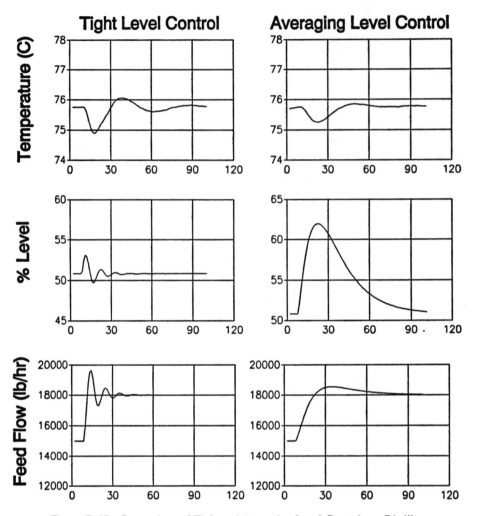

Figure 7–15   Comparison of Tight and Averaging Level Controls on Distillation Tray Temperature Variability

Level control loop tuning can easily be calculated by specifying a few parameters.

$$Kc = \frac{2\,H_{\max}}{\tau_c\,F_{\max}} \tag{7–4}$$

TABLE 7–7 COMPARISON OF TUNING PERFORMANCE

	Tight Level Tuning	Averaging Level Tuning
Temperature Deviation	1 degree	0.5 degree
Temperature Standard Deviation	0.26 degrees	0.17 degrees
Feed Flow Overshoot	9%	3%
Level Deviation	3%	12%

$$Ti = \frac{Kc \, (\zeta \tau_c)^2 \, F_{max}}{H_{max}} \qquad\qquad (7\text{--}5)$$

where

H_{max} = maximum holdup in pounds between the level taps
F_{max} = maximum manipulated flow in lbs/min
τ_c = closed loop time constant in minutes
ζ = damping ratio

The larger the closed-loop time constant, the smaller the controller gain and the slower the response. Damping ratio has to do with the number of oscillations in the response. For averaging level control the damping ratio should be 1.0 or greater. The tuning should be set so the level stays in the tank for the maximum upset.

There are even more complex control algorithms to maximize smoothing. Nonlinear controllers make the controller gain a function of the deviation in level. Control action is very sluggish around the setpoint, and the action accelerates the further the level is from the setpoint [Shunta 1976].

MULTIVARIABLE CONTROLS

A major source of variability in chemical processes is the interaction among control loops on the same or adjacent equipment. Each control loop tries to keep its process variable on aim, but because manipulative variables can often affect more than one process variable, it causes other loops to vary. The problem is particularly bad when the speed of response of the variables is about the same. The problem gets worse the tighter the controllers are tuned, and it is not uncommon for instability to occur. One example of interaction was covered in the section on averaging level control. Another classic case is when the purities of two product streams on a distillation column are tightly controlled. The first thing we try to do to minimize the interactions is to de-tune the least important controller so it does not react very quickly to upsets. If both products are equally critical and have to be tightly controlled, we have to employ more sophisticated multivariable controls to compensate for interactions.

There are a couple of ways to design multivariable controls. The first is like feedforward control. Referring to Figure 7–8, the ratio control loop is actually an interaction compensator. It adjusts the B stream flow in anticipation of the variability in concentration that will be caused by upsets in the A stream flow. Figure 7–16 shows how this type of *interaction compensation* can be implemented on a distillation column.

The top temperature is controlled by manipulating the reflux flow, and the lower temperature is controlled by adjusting the steam flow to generate boilup in the reboiler. Changing reflux will affect both the top and bottom temperatures, and so will changing boilup. The interaction compensators, or decouplers D1 and D2, make compensatory adjustments to the other manipulative variable to counteract the upset.

The decouplers are calculated from models relating the changes in top and bottom temperatures to changes in steam and reflux flows.

$$\Delta T_{top} = G_1 \, \Delta \, \text{Reflux} + G_2 \, \Delta \, \text{Steam} \qquad (7\text{–}6)$$

$$\Delta T_{bot} = G_3 \, \Delta \, \text{Reflux} + G_4 \, \Delta \, \text{Steam} \qquad (7\text{–}7)$$

where the Gs are open-loop dynamic response models (typically lags and deadtimes) for the temperatures. The decoupler for the top temperature is solved from Equation 7–8 by setting ΔT_{top} to zero, which says that the temperature will not change.

$$D1 = -\frac{G_2}{G_1} \qquad (7\text{–}8)$$

Likewise, $D2$ is calculated from Equation 7–9.

$$D2 = -\frac{G_3}{G_4}$$

The decouplers are implemented in the DCS from lead-lag and deadtime functions. They work well for simple 2×2 systems. Getting more complex, par-

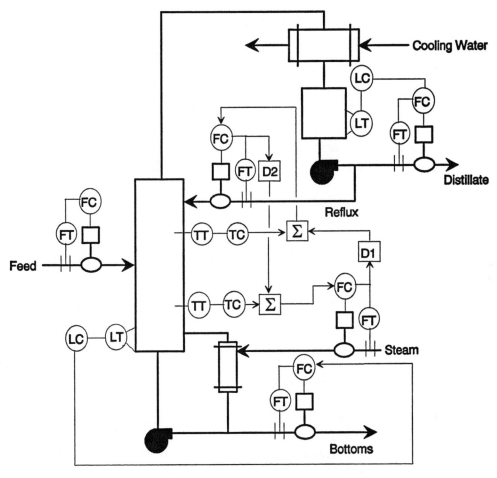

Figure 7–16   Decoupling Controls on a Distillation Column

ticularly when deadtime becomes significant, starts to make design of decouplers increasingly difficult using single-loop controllers and calculating functions.

MODEL PREDICTIVE CONTROL

For the more difficult and complex multivariable systems, *model predictive control* (MPC) should be considered. Complex multivariable systems, particularly with deadtime, are intractable with simple functions and controllers. In the U.S.,

MPC was developed for the petroleum refining industry by engineers at Shell Oil [Cutler 1979]. Since then, MPC has become a somewhat standard control strategy for FCC units and other multivariable refinery operations. These units are characterized by the following:

- Multivariable systems with several key controlled and manipulative variables
- With significant interactions among the variables
- With problematic dynamic behavior like long deadtimes and time constants, and inverse response
- Requiring tight composition control
- With a large throughput so small-percentage improvements translate into large economic benefits
- With inherent nonlinearities where the process gains and dynamics change depending on operating conditions
- Subject to process and equipment constraints

In the CPI, reactors and distillation columns represent the greatest potential for economic benefits, so they would logically be the prime candidates for MPC. Gerstler (1991) discusses MPC applied to several distillation columns in an olefin plant. They reported significant increases in throughput and yield, and reductions in variability of product properties. However, they caution that MPC is not the answer to every control problem; but for complex equipment, MPC has proven to dramatically improve control.

Another critically important unit operation in chemical processing is reaction. Although there are not as many reports in the literature of successful applications of MPC to reactors, we know that some very significant improvements have been made in reactor control with this technology. Lines (1993) applied MPC to a multivariable polymer reactor unit that displayed process nonlinearities and significant interaction among the process variables. The business drivers in this case were to reduce transition times and off-specification products when going from one product grade to another. MPC achieved a 60% reduction in transition time between most product grades and a 50% reduction in off-standard product, plus a 15% improvement in first-pass yield. Some of the variables they measured had significant deadtime, so they used linear estimation algorithms to infer the variables from current reactor conditions. We will cover this subject in the next chapter.

MPC has some real advantages in the appropriate application, but as already stated, it should not be used across the board because it is complex compared to conventional DCS controls. MPC is a software-based control technique that incorporates linear time response models. Some versions are implemented in a host computer and some available in a DCS. The model consists of a matrix of coefficients where each coefficient is the change in the process variable at successive time steps to a unit change in the manipulative variables. The model has two purposes: it predicts the response of the process variables several steps into the future, and computes present and future changes in the manipulative variables to bring the process variables back to setpoint. It determines the control actions by minimizing the deviations between the predicted process variables and the setpoint by a constrained least-squares optimization.

The two most common versions of model predictive control are called IDCOM (identification and command) and DMC (dynamic matrix control). The methods are basically the same, although one uses a step response model and the other an impulse response model. Another difference is in the way the controllers are tuned. The principle can be illustrated by a single-input, single-output process. Using IDCOM as an example, the predicted response of the process variable X from an impulse response model is given by Equation 7–10 [Martin 1981].

$$X(k+N) = \sum_{i=1}^{N} h_{N-i+1} \, \Delta M(k+i-1) \qquad (7\text{--}10)$$

where

h = impulse response coefficients, the change in X for a unit change in M
ΔM = change in the manipulative variable
N = number of steps into the future
k = current time

For a multivariable system, the responses are expressed in matrix notation. The predicted values of X are compared to a specified reference trajectory that returns X to the setpoint.

$$X_R\,(i+1) = \beta\, X_R\,(i) + (1-\beta)\, X_{SP}$$

where

X_R = desired exponential trajectory
X_{SP} = setpoint
β = tuning factor for the exponential trajectory

The algorithm uses the impulse response model to calculate a set of future moves of the manipulative variable that minimizes the deviation between X and X_R several steps into the future. Only the first move of the manipulative variable is made, and the process is repeated. Tuning is accomplished by specifying the slope of X_R.

The DMC algorithm uses a similar strategy, only with a step response model. It calculates the manipulative variable adjustment that minimizes the sum of squared deviations between the predicted process variable trajectory and the setpoint (i.e., the process variance). Since minimizing only the process variance can lead to very large adjustments in the manipulative variable, the objective function contains both the process variance and the manipulative variable variance so that the user can specify what is more important: to minimize the process variance or make small manipulative variance adjustments. Tuning is done by adjusting a weighting factor on the manipulative variable term.

The strengths of these controllers are that they contain the responses of each process variable to each manipulative variable so all the interactions are compensated for and feedforward control implicitly implemented. In addition, deadtime is modeled and compensated for. If done properly, MPC should take care of all variability except that due to some random and unmeasured disturbances. A large part of the implementation effort is to identify the response models from plant tests. The next chapter shows how other types of models, such as neural networks, are being used in place of step response models.

Over time, the linear models may need updating. For very nonlinear processes where the gains and time constants change with process conditions, it may even be possible to adjust the models on-line. Another approach is to have more than one version implemented so that the right one can be used for the appropriate steady-state condition. Currently, researchers are developing nonlinear model predictive controls. Georgiou (1988) presents a version of MPC that uses a nonlinear transformation (logarithm) on the process variable. Ogunnaike (1986) shows how DMC can be expanded for singlevariable systems with time-varying time delays and steady-state gains.

LOGIC AND AUTOMATIC CONTROLS

A large source of yield losses is off-specification material generated during startups, shutdowns, rate changes, and product transitions. This is when not only process conditions are varying, but the process variables are outside of the standard operating conditions required for producing acceptable product. We would like the control system to do two things during these periods: minimize variability and get to standard operating conditions as quickly as possible.

Variability can be reduced somewhat by having optimized operating procedures for the operator to follow. However, even with good procedures, operating personnel are not always consistent in following them. Automation can provide an additional degree of consistency in getting to the desired operating state, but unless the automation logic has been optimized for minimum off-specification product, you still will not achieve the best yield. In addition, inconsistent operation can still occur as a result of unmeasured disturbances. Thus, the automatic controls must be well-tuned and designed properly. The other benefit of automation, besides providing consistency, is increased speed in getting to the proper state, which impacts the capacity of the unit. In a sold-out condition, this is where large economic benefits can be realized.

The combination of logic and automatic controls may not be as valuable for single-line continuous processes as they are for multiproduct continuous or semi-continuous operations like some polymer lines. In one polymer process, automating the rate changes achieved a sustained yield improvement of 2–4%. It did this by coordinating the setpoint changes to a number of control loops to keep the properties of the polymer as constant as possible. The rate of return of the automation system was 25:1.

In another example, the challenge was to minimize the time in getting from one product grade to another by properly changing the operating parameters. This scheme used an optimization model to calculate the best set of process variable moves from one steady-state condition to the other. The objective function was maximum profit by taking into account yield losses versus transition times. Going a step further, some plants have optimized the selection of the next product based on requirements for filling orders and minimizing transition losses.

While transition controls automate the change from one continuous operating state to another, batch controls automate the startup and shutdown and control the steady-state operation of the process. Batch controls are a combination of logic and automatic controls. The logic controls open and close valves and initiate and terminate control actions. Automatic controls keep the process at the setpoint conditions. A typical case in point is shown in Figure 7–17. The operation is as follows:

- Precharge the reactor with one reactant.
- Heat up the contents.
- Add the second reactant to start the reaction.

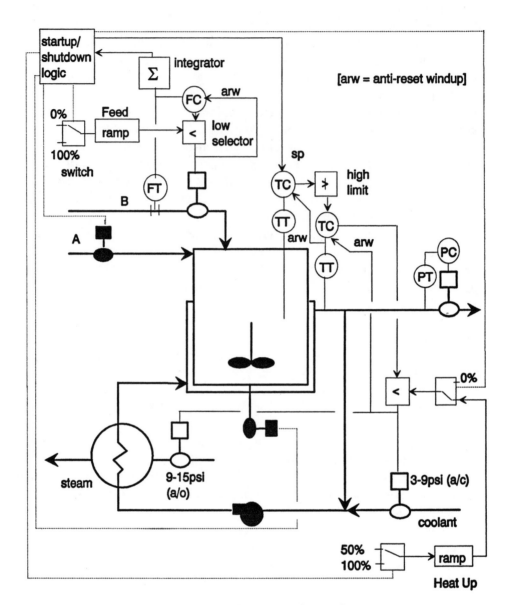

Figure 7–17   Batch Controls on a Reactor System

- Maintain temperature control until the reaction is complete.
- Cool the reactor contents.
- Discharge product.

The logic controls precharge the first reactant. Then logic opens the steam valve to heat up the contents to a preset temperature on the temperature controller. The automatic reactor temperature controller takes control of the steam and cooling water valves to maintain the temperature. Then the logic starts ramping open the feed until the feed flow controller takes over and maintains the flow at a preset rate until the proper mass of material has been added; then it closes the feed valve. The reaction is exothermic, so as the reaction takes place, heat is generated and the temperature control maintains the correct temperature by bringing on more cooling. The temperature controls are cascaded to speed up the response and compensate for any upsets originating in the jacket steam or cooling water systems. Note that there is a high limit on the jacket temperature controller setpoint, so no hot spots can occur in the jacket. The controllers also need to be equipped with anti-reset windup logic to prevent them from saturating when they are being overridden. When the reaction is complete, the logic puts on full cooling to achieve the desired discharge temperature. Finally, the contents are discharged to storage. This type of logic is configurable in modern distributed control systems and in programmable logic controllers because both devices have logic and continuous control algorithms and functions.

One industrial example of batch controls was able to raise particle size distribution Ppk from 1.0 to 3.1 through automating the steps. The batch controls were all implemented in a DCS. Batch supervisor software contained a database of all the operating parameters, setpoints, temperatures, sequence times, and so forth for making several grades of product. The parameters were downloaded to the control and sequence logic, which executed the control actions per the specified parameters. One critical action was ramping reactor temperature at a prescribed rate to minimize the distribution of particle size.

In addition to transitioning and batch controls, the combination of logic and automatic controls is used to start up and shut down even continuous processes. This is done for a variety of reasons.

- Process equipment is used for different purposes; thus, startups and shutdowns occur frequently enough to make the additional complexity worthwhile. One example is a solvent recovery system consisting of one or more distillation columns that purify and recycle solvents for a reaction process. The reaction system produces several grades of product using different solvents. Often, the distillation equipment can be used to recover more than one solvent, but the columns must be shut down and cleaned out between each campaign.

- The most hazardous time during the operation of chemical processing equipment is during startups and shutdowns, when variables are outside of normal operating limits and a lot of different things are happening all at once. Automating these steps guards against inadvertently making the wrong moves, which may lead to unsafe conditions.
- When product is sold out or the production schedule under time constraints, automating the steps may speed up startup or shutdown. It is not unusual to gain savings of thousands of dollars per hour by getting on-line smoothly and quickly.
- When equipment and process configurations are complex, startup and shutdown automation may really be necessary to get on-line in a timely manner.

The logic for automating startups and shutdowns can be implemented by using a combination of rather simple functions that are standard modules in a DCS. The main effort is in thinking through each step and deciding what things need to occur and in what sequence. Once that is done, the implementation is not very difficult. The next example shows the kinds of logic functions and implementation required to start up and shut down a distillation column.

The complete configuration is shown in Figure 7–18. It looks awesome at first, until you realize the strategy consists of a number of simple functions applied in a variety of ways. The functions are:

- High and low signal selectors ($>$, $<$), which select and transmit the highest or lowest signals from among its inputs
- A signal scaler $f(x)$, which is no more than an equation for a straight line [$f(x) = m\,x + b$]
- A ramp generator whose output increases slowly according to a specified ramp time and decreases rapidly so a valve can be opened slowly and closed quickly
- On-off switches used to change operating modes

In the example, the operator only has to operate two switches to start up the column: the startup/shutdown switch, which opens the feed and steam, and the total reflux-normal switch, which operates the distillate and bottom valves.

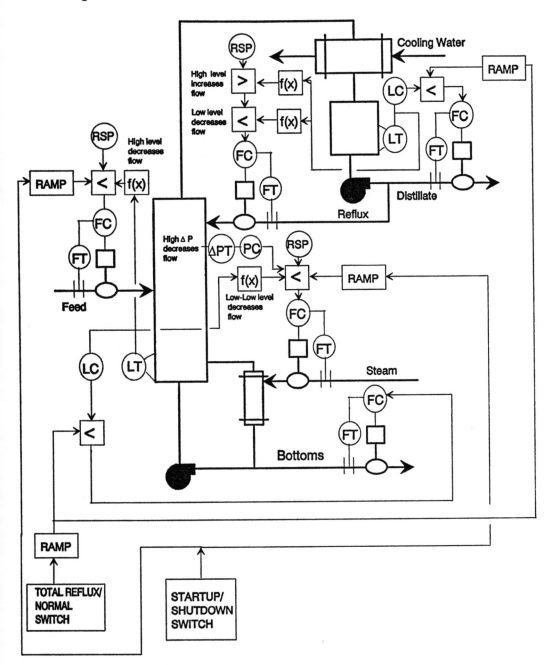

Figure 7–18 An Example of Startup/Shutdown Logic for a Distillation Column

The logic for a startup works as follows:

- The startup/shutdown switch is initially in the shutdown mode, so the feed and steam flows are off. The total reflux/normal switch is initially in the total reflux mode, closing the distillate and bottom so nothing is taken out of the column while it is being inventoried and heated up.

- Although this logic is connected to the setpoints of the flow controllers, it could just as easily operate the valves directly. If that were the case, a means would have to be provided to bypass the signal selectors so that the operator could always get to the valves directly. In this situation, the operator can put the flow controls in local automatic to get direct control.

- The controllers are in full automatic, either in cascade or local automatic as the case may be. The remote setpoints (RSP) are set to the normal operating values. Just as in the reactor example shown previously, the PID controllers whose output can be blocked from adjusting the valve, such as in the case where the controller output feeds a signal selector, need to have external reset feedback. Otherwise, if the controller signal is not selected, the integral mode will continue to integrate the deviation from setpoint and saturate at either 100% or 0%, depending on the action of the controller. External reset feedback tracks the feedback signal and does not saturate. This is not a problem for proportional-only controllers.

- The cooling water to the condenser must be on.

- The startup-shutdown switch is placed in the startup position, and the startup signal increases the feed flow controller setpoint according to the ramp rate specified in the configuration. When the ramped signal increases above the remote setpoint, the RSP signal will be selected. The steam setpoint will not be allowed to increase as long as no level is indicated by the low-low base level override. The low-low level override operates nominally between 0% and 10% level, according to the way $f(x)$ is scaled. Assuming the level range is 0–100%:

$$f(x) = 10 \left[\frac{\text{flow span}}{100\%} \right] [\text{level (\%)}] \qquad (7\text{–}12)$$

When the level is equal to or less than 0%, $f(x) = 0$ and the steam is closed. Above 0%, $f(x)$ starts to increase and will reach a maximum at 10% level. This ensures that steam is not introduced to an empty reboiler.

If, at any time, the base level gets above nominally 75%, the high base level override will start to decrease the feed. Again, it depends on how the high level override is scaled. In this case $f(x)$ is

$$f(x) = -4\left[\frac{\text{flow span}}{100\%}\right] [\text{level} (\%)] + 4 [\text{flow span}] \qquad (7\text{--}13)$$

- The level in the base will start to increase, and when a level is indicated, the low-low base level override will allow the steam flow to increase. Vapor boilup will be generated, and condensate will start to fill the condensate tank. When the condensate level gets above 0%, the low tank level override will allow the reflux flow setpoint to increase until it reaches the remote setpoint. The low tank level override is calibrated as follows:

$$f(x) = 4\left[\frac{\text{flow span}}{100\%}\right] [\text{level} (\%)] \qquad (7\text{--}14)$$

Thus, when the level is above 25%, the reflux override is not active. If, for some reason, the level gets above about 75%, a high level override will force the reflux to increase and control level. That override is calibrated as:

$$f(x) = 4\left[\frac{\text{flow span}}{100\%}\right] [\text{level} (\%)] - 3 [\text{flow span}] \qquad (7\text{--}15)$$

The overrides are simply proportional-only level controllers that only come into play when the level gets beyond *soft* constraints. It is assumed that during the startup phase, pumps will be running when they are needed.

- When the column is heated up and vapor and liquid rates are established, the operator switches to *normal* mode, and the distillate and bottom flows are ramped up until taken over by the normal level controllers.
- Any time during the startup or normal operation, if the column pressure drop exceeds the maximum setpoint indicating incipient flooding, the high ΔP override will throttle back on the steam. This override is implemented with a PI controller instead of a scaler function because we want to control at the maximum ΔP setpoint, whatever steam flow that requires. With a scaler (proportional-only) override you have to know what steam flow is required at a given ΔP in order to calibrate it. We do not know that relationship *a priori*.
- In order to shut down, the operator could switch to *shutdown*, which would cut off the feed and steam and let the level controllers de-inventory the column, or switch to *total reflux*, which would shut off the distillate and bottom flows and let the high base level override shut off feed. Also, the operator could shut off the feed independently.

This example illustrates that startup/shutdown logic can be implemented with fairly simply functions coupled with continuous controllers. Since the functions can all be implemented in the DCS, the capital investment is very little if any at all, which makes it attractive.

FUZZY LOGIC CONTROLS

A new control technique appearing more frequently in the literature is based on fuzzy logic (FL). FL comes from set theory pioneered by Zadeh (1965). Vom Berg (1993) describes FL as a rule-based artificial intelligence algorithm that uses human-like reasoning to determine control actions. FL-based controllers are targeted for cases in which PID does not work well, such as multivariable processes with significant deadtime, which generally require some type of model-based controls. Conventional mathematical algorithms require exact numerical equations, whereas FL control is based on approximate linguistic descriptions to specify control actions. A typical control rule may look like:

IF (PRESSURE = HIGH .AND. TEMPERATURE = MEDIUM HIGH) THEN . . . take such and such control action. The control action then depends on the degree that pressure and temperature are high rather than on fixed values. You can see that the rules are similar to human reasoning.

In many complex dynamic processes, a mathematical model is difficult to develop; so FL offers a somewhat simpler approach, at least conceptually. Roffel (1991) has a good description of how the FL rules work as applied to control of a polymerization reactor. FL was chosen because developing an adequate conventional control strategy would have been prohibitively expensive. They found that the FL controls reduced the standard deviation of the key product property by about 40%.

Vom Berg (1993) promotes FL for temperature control in terms of augmenting a PID controller. FL can modify the controller tuning and cause the controller to have a faster speed of response. Several FL controllers are now on the market.

As time goes on, fuzzy logic will certainly become more prevalent in control systems [Samdami 1993]. At this point, it is still very much a curiosity.

COMBINING STATISTICAL AND AUTOMATIC PROCESS CONTROL FOR QUALITY

It has been shown throughout this book that statistics and process control work together in improving the manufacturing operation. The integration of these two technologies has become increasingly important as more emphasis is placed on product quality to stay competitive. However, there is a large gap between how applied statisticians and process control engineers approach quality control. Narrowing this gap will result in being more effective in producing high-quality products.

MacGregor (1988) correctly points out that in the chemical processing industries, process control engineers, who are predominantly chemical engineers by education, have not had the statistical training necessary to use noisy and infrequent data (which characterizes what we get from quality control labs) in control strategies. The task of quality control has therefore been the domain of applied statisticians in most companies. On the other hand, applied statisticians have not had the training in the analysis and modeling of dynamic processes or continuous automatic process control of process control engineers. As a consequence, statistical control methods have often been incorrectly applied for on-line quality control. This unfortunate state of affairs has led to a large gap in the technology on how to apply infrequent quality data in continuous control strategies. MacGregor (1988) and his colleagues have done a real service to industry by promoting the idea that statistics and process control need to be viewed as complementary technologies for quality control.

SPC is based on the theory that the process mean is constant, and the product or process data are independent and identically distributed, which is not always the case in chemical process control applications. It is likely that the data are not independent but autocorrelated (data influence by previous data), and the process mean drifts. Of course, the more infrequently the data are measured, the more independent the data are likely to be.

MacGregor (1988) points out that depending on the type of process, SPC or APC algorithms may be appropriate for control. For example, if the process has no dynamics, that is, the process reaches a new steady state by the time the next data is received (what control engineers call gain-only) but there is a cost associated with each adjustment, a traditional SPC algorithm such as CUSUM or EWMA may be optimal for control. This typifies discrete parts manufacturing. For most chemical processes, however, the cost of making an adjustment is nil, so there is no reason not to make an adjustment every time a new data point is received. For a process with no dynamics, an integral-only algorithm is appropriate. One example is a control loop where the measurement is a chemical analysis from the control lab. An integral-only algorithm has the following form:

$$M_n = \frac{Ts}{T_i}(X_{sp} - X_n) + M_{n-1} \qquad (7\text{--}16)$$

where

M_n = controller output
M_{n-1} = previous output
Ts = sampling period
T_i = Integral time, minutes/repeat
X_{sp} = setpoint

For control loops where the measurement is frequent enough that it is influenced by the previous measurements, a dynamic control algorithm such as a PI, PID, or more complex ones are appropriate (see Equation 7–1). This is the most common scenario in chemical processes because of large in-process lags caused by tanks and recycles. Table 7–8 summarizes how the various algorithms are applied.

Infrequent measurements may at first glance seem to have no dynamics; i.e. the process response reaches steady state between samples. However, for some measurements like composition, there is a possibility that there are long-term, slow dynamics present because of recycles and large process storage vessels. If the process behaves like an integrating process (non-self-regulating), an integral-only algorithm would cycle and may even be unstable over the long term. Therefore, it may be prudent to use a control algorithm that has proportional action as well as integral action.

Chapter 9 will illustrate how SPC and APC are combined so that the monitoring capabilities of SPC are used effectively to augment the compensatory function of APC.

ADAPTING CONTROLS TO CHANGING OPERATING CONDITIONS

This section discusses how controls are adapted on-line to maintain a desired level of stability. *Adaptive control* is important because it is one of the ways to compensate for the special-cause variability emanating from changing process

TABLE 7–8   APPLICATION OF CONTROL ALGORITHMS

Process/Measurement	Cost of Adjustment	Optimal Control Algorithm
No Dynamics/Infrequent	Yes	SPC: EWMA, CUSUM, SHEWHART
No Dynamics/Infrequent	None	Integral - Only
Dynamics/Frequent	None	PI/PID/other

dynamics. There have been many papers written on this subject, but not all the techniques have made their way into common usage in the CPI. One of the reasons is that it is difficult to make the controllers smart enough to be able to distinguish between real changing process dynamics and process noise.

The PID algorithm discussed earlier contains fixed tuning parameters that are selected to achieve a desirable dynamic response at some operating condition. Seldom do processes exhibit the same dynamic characteristics under all operating conditions. Therefore, if operating conditions change, the closed-loop response may not be the same as before unless the tuning is adapted to the new conditions. With some processes, the dynamic response can change so dramatically that it requires a different set of tuning parameters to even maintain stable operation. PH control is a prime example.

Often the controller gain is the critical parameter that needs to be adapted. A technique called *gain scheduling* automatically adjusts the controller gain based on some measurable operating condition (such as flow rate) to maintain the same dynamic response or degree of stability. Gain scheduling is an easy form of adaptive control to implement in the DCS. Of course, *a priori* knowledge of the relationship between the required gain and the measured condition is required before gain scheduling can be implemented.

More recently, single-loop digital controllers and DCS control algorithms have been developed containing software that identifies the process response

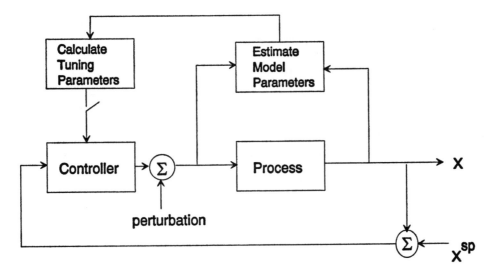

Figure 7–19 An Adaptive Control Strategy

characteristics and calculates the appropriate tuning parameters from these data automatically. A block diagram of the strategy is shown in Figure 7–19. Many of these self-tuners introduce small perturbations into the process from which the process characteristics can be identified. Often the user has the option of calling up the self-tuner on demand or leaving it in the automatic tuning mode continuously. Except for some highly nonlinear processes that require frequent retuning, most practitioners would agree that self-tuning on demand is safer because it is a more controlled environment.

None of these methods are fool proof, but still they have come a long way towards improving process control. Since keeping control loops properly tuned is a big step towards maintaining the benefits from process control, serious consideration should be given to having self-tuning capability available to the operator.

REFERENCES

BUCKLEY, P. S., LUYBEN, W. L., and SHUNTA, J. P. *Design of Distillation Column Control Systems*. Research Triangle Park, NC: Instrument Society of America, 1985.

CUTLER, C. R. and RAMAKER, B. L. Paper 51b, AIChE 86th National Meeting, Los Angeles, 1979.

DESHPANDE, P. B. and ASH, R. H. *Computer Process Control*. Research Triangle Park, NC: Instrument Society of America, 1988.

GEORGIOU, A., GEORGAKIS, C., and LUYBEN, W. L. Nonlinear Dynamic Matrix Control for High Purity Distillation Columns. *AIChE Journal*, Vol. 34, No. 8, pp. 1287–1298, August, 1988.

GERSTLE, J. G., and HOKANSON, D. A. Opportunities for Dynamic Matrix Control in Olefins Plants. Paper presented at 1991 AIChE Spring Meeting, Houston, Texas, April 7–11, 1991.

HARRIS, T. J., MACGREGOR, J. F., and WRIGHT, J. D. An Overview of Discrete Stochastic Controllers: Generalized PID Algorithms with Dead-Time Compensation. *The Canadian Journal of Chemical Engineering*, Vol. 60, No. 3, pp. 425–432, 1982.

LINES, B. et al. Polyethylene Reactor Modeling and Control Design. *Hydrocarbon Processing*, pp. 119–124, June 1993.

LIPTAK, B. Controlling and Optimizing Chemical Reactors. *Chemical Engineering*, pp. 69–81, May 26, 1986.

LUYBEN, W. L. *Processing Modeling, Simulation, and Control for Chemical Engineers*. New York: McGraw-Hill, 1990.

MacGregor, J. F. On-Line Statistical Process Control. *Chemical Engineering Process,* pp. 21–31, October 1988.

Martin, G. D. Long-Range Predictive Control. *AICHE Journal,* Vol. 27, No. 5, pp. 748–753, September 1981.

McMillan, G. K. *Tuning and Control Loop Performance.* Research Triangle Park, NC: Instrument Society of America, 1990.

Meyer, C., Seborg, D. E., and Wood, R. K. A Comparison of the Smith Predictor and Conventional Feedback Control. *Chemical Engineering Science,* Vol. 31, pp. 775–778, 1976.

Ogunnaike, B. A., and Adewale, K. E. P. Dynamic Matrix Control for Process Systems with Time Varying Parameters. *Chem. Eng. Commun.,* Vol. 47, pp. 295–314, 1986.

Roffel, B., and Chin, P. A. Fuzzy control of a polymerization reactor. *Hydrocarbon Processing,* pp. 47–49, June 1991.

Samdani, G. Fuzzy Logic—More than a Play on Words. *Chemical Engineering,* pp. 30–33, February 1993.

Schleck, J. R., and Hanesian, D. An Evaluation of the Smith Linear Predictor Technique for Controlling Deadtime Dominated Processes. *ISA Transactions,* Vol. 17, No. 4, pp. 39–46, 1978.

Shinskey, F. G. *Process Control Systems.* New York: McGraw-Hill, 1979.

Shunta, J. P. A Compensator for Inverse Response. Paper presented at ISA/84 International Conference and Exhibit, Houston, Texas, October 1984.

Shunta, J. P., and Fehervari, W. Nonlinear Control of Liquid Level. *Instrumentation Technology,* Vol. 23, No. 43, January 1976.

Smith, C. L. *Digital Computer Process Control.* Scranton, PA: Intext, 1972.

Smith, O. J. M. A Controller to Overcome Dead Time. *ISA Journal,* Vol. 6, No. 2, 1959.

vom Berg, H. Fuzzy Logic: a clear choice for temperature control. *I & CS,* pp. 39–41, June 1993.

Zadeh, L. A. Fuzzy Sets. *Information and Control,* Vol. 8, pp. 338–353, 1965.

Ziegler, J. G., and Nichols, N. B. Optimum Settings for Automatic Controllers. *Trans. ASME,* Vol. 64, p. 759, 1942.

8

Inferential
Measurements

Chapter 7 showed that the opportunity to reduce variability becomes smaller as the measurement frequency decreases. This usually becomes an issue with quality variables or product properties because we often have to rely on the laboratory analyses or on analytical equipment that operates on cycles of several minutes. If the minimum variance we can achieve by improved feedback control for the given measurement is not sufficient, there are some options:

- Supplement the feedback strategy with feedforward or cascade control. This depends on being able to make the necessary additional measurements.
- Install a measuring device that operates on a faster cycle time. This may or may not be feasible.
- Infer or estimate the variable by some mathematical technique.

The last option is the subject of this chapter. Since this is a very complex subject, it is difficult to do it justice in one chapter. Therefore, the scope of this chapter will be limited to providing a brief overview of some topics related to process control.

ESTIMATING VARIABLES

Ray (1981) points out that we can view the problem of estimating a variable in terms of the time frame. Estimating the current value from current and past data is called *filtering,* and estimating future values is called *prediction.* Inferential measurements have to do with making current estimates of variables from current and past values of one or more other measured variables.

There are a number of approaches for estimating the current value of a variable. One is to calculate the variable from a mathematical model describing the physics and chemistry of the process (first principles model). The inputs to the model are the measured process variables such as flow, temperature, pressure, and concentration. Another approach is to develop a purely statistical model from the available measurements. These models have no discernible relationship to the actual physical nature of the process, which in some ways is a disadvantage. A third method combines a physical or empirical model with statistical techniques to estimate process variables in the face of noisy measurements and random disturbances. A well-known example of this is the Kalman filter [Kalman 1960]. Wells (1971) defines the filter this way: "The discrete Kalman filter equations specify an optimal estimate of the state of a linear, time-varying, dynamic system observed sequentially in the presence of additive white Gaussian noise." Since the measurements are noisy, the statistical techniques are employed to deal with them. Wells (1971) reports on an application of the Kalman filter to estimate the unmeasured output variables (states) of a chemical reactor.

PROCESS MODELS

An example of a simple process model consists of an algebraic equation that defines the physical relationship between the variable of interest and one or two other measured variables. For example, the composition of a binary (two-component) liquid in a distillation column can be deduced from the temperature in the column at a known pressure. Since pressure fluctuates and causes the temperature to change, the effect of pressure is first subtracted from the measured temperature so that the corrected temperature reflects only the composition at that location.

$$T_{corr} = T_{meas} + \frac{\delta T}{\delta P}(P - \bar{P}) \qquad (8\text{--}1)$$

where

T_{corr} = temperature that predicts composition

T_{meas} = measured temperature

$\dfrac{\delta T}{\delta P}$ = change in temperature per unit change in pressure (slope of vapor
pressure curve)

$\bar{P}$ = nominal steady-state pressure

P = measured pressure

Controlling the corrected temperature is effectively controlling composition. There are some shortcomings using this relationship, however. We do not always have a binary mixture, and temperature is not always measured at the point in the column where we want to control the composition. Then there is no longer a one-to-one correspondence between temperature and product composition.

Sometimes simple regression equations are used to infer a variable. For example, the composition has been shown to be a nonlinear function of temperature in the column.

$$\text{Log } X = a_0 + a_1 T \qquad (8\text{--}2)$$

where X is the composition and T the corrected temperature from the previous equation. This statistical model is developed from observed temperature and composition data as opposed to modeling the physical system.

An application was recently reported by Dollar (1993) in which simple models were used to predict composition of top and bottom products and flooding in a distillation column. The model-based controls successfully prevented large variations that had been experienced before the controls were implemented. That made it possible to maintain tight control of compositions.

Sometimes we have to use more complex mechanistic or first principles models for prediction. These comprise differential and algebraic equations that describe the material and energy balances and reaction kinetics of the physical process. The equation parameters have to be calibrated to reflect the actual process. The advantage of these models is that they include the nonlinearities and predict the actual dynamic response. On the other hand, there may be many parameters that have to be adjusted as process conditions change to keep the outputs valid. It is particularly critical that the dynamics of the model match the process dynamics closely, or the predicted variables will not be synchronized with the process. If the model output is used for control, the results will be poor. One way to calibrate the model is with actual lab measurements. The calibration compensates for modeling errors, nonlinearities, measurement errors, and the effect of unmeasured disturbances. The lab measurements should first be filtered to remove noise or they may compound the problem. Ardell (1983) reports an application of this technique for reactor control.

In general, on-line dynamic first principles models for control purposes have had limited utility and success. Sometimes the physics of the process are just too complicated to model adequately. Models may also fail to be used because of the day-to-day support required to keep them calibrated. Then there is the problem of not having reliable and accurate measurements. The measurements must be routinely reconciled or validated to make sure that no erroneous input data to the model will throw the results off. The same caution must be taken with operator inputs. Friedman (1992) states that it has been shown statistically that one out of 50 manually entered inputs will be incorrect. It is imperative, therefore, that inputs are validated against an accurate standard, by material balance, or through some other appropriate technique. The conclusion most experienced people reach about models is that if physical models must be used, it is best to keep them simple.

STATISTICAL MODELS

One option to dynamic first principles models are time series models developed from observed data. The key is to have an adequate number of reliable measurements. Undetected sensor failures, uncalibrated or mislocated sensors, and inadequate data storage all present problems for statistical models. Common time series models are linear autoregressive moving average (ARMA) models [Harris 1982]. Since these models have their roots in statistics, process control engineers in general have not dealt much with them. This is unfortunate because they address random disturbances, which are very much present in chemical processes.

Other types of statistically based techniques have been promoted over the years to estimate variables. *Parameter estimation* estimates the parameters of a specified model structure (such as a first-order lag with deadtime) from observed process input and output data. The models are then used for estimating process outputs. Even though these techniques have been around for some time, they still require a fair amount of expertise to make them work successfully.

More recently, *chemometrics* techniques and *neural networks* have been introduced to the chemical industry. Chemometrics uses multivariate linear regression techniques such as *partial least squares* or *projection to latent structures* (PLS) and *principle component regression* (PCR) to build static linear estimators. They are particularly effective with large numbers of highly correlated variables that pose problems for traditional multiple linear regression techniques. They reduce the dimensionality of the data set while not eliminating useful information, and also reduce the effect of noise. Kosanovich (1991) reports on some research in which these methods allowed them to reduce the dimensionality of their system from 14 inputs to three. The models they developed were to determine predictions of bottom composition in a distillation column.

Mejdell (1991) reports on a successful application of PLS to estimate product compositions in a pilot-size high-purity distillation column. A simple linear estimator expresses the relationship between compositions and temperatures as follows:

$$
\begin{aligned}
X_D &= g_{11} T_1 + g_{12} T_2 + \dots \\
X_B &= g_{21} T_1 + g_{22} T_2 + \dots
\end{aligned}
\tag{8-3}
$$

where

X_D = deviation of distillate product composition
X_B = deviation of bottom product composition
g_{ij} = process gains
T_i = temperatures

However, in the case they studied, the temperatures were correlated, and there were not enough degrees of freedom to solve for the gains. Instead, they used PLS to reduce the temperature data into a fewer number of independent latent variables. Then a relationship was found between the compositions and the latent variables. They found that estimators based on actual operating data gave very good results. Just as with all regression techniques, it is critical that the data set used to develop the estimator capture all the expected variations in the input variables. Otherwise, the resulting estimator will not be accurate in all situations, because regression techniques are not good for extrapolation. This may mean a large investment in time to collect enough data to model complex processes. You could use a simulation to generate the required data, but Mejdell (1991) found that the results were not as good as with experimental data. Other good references on applying PLS to the chemical industry are Kresta (1990) and Piovoso (1992, 1993).

A recent static estimation method that generates nonlinear estimates is called *neural networks*. An advantage of this method is that it does not require that the model structure be specified beforehand as in other techniques. Instead it develops a structure, an internal network of weighted signals, by training itself on a set of observed input and output data. The trained model can then be used to predict outputs based on a new set of inputs. Again, as with regression techniques, a large range of data must be used for training because the estimator cannot accurately predict a pattern on which it has not been trained. This can turn out to require a large investment in time.

Figure 8–1 shows the structure of a neural network system. Normalized operating data are input to the first layer of nodes. The outputs from the nodes (*O*) are weighted (*W*) and propagated through the network of other nodes to predict the plant outputs. The nodes are arranged in layers. The number of layers is defined by the modeler to get the best results. The individually weighted signals are summed together in each node. The sum passes through a nonlinear normalizing function so that the output takes on values between zero and one. The signals continue to propagate through the various layers of nodes until the final outputs are calculated. The final outputs are scaled to the correct engineering units and compared to the plant output data. The difference is used to readjust the weighting factors. The process is repeated until the network produces the desired outputs. One disadvantage of neural net models is that the model parameters cannot be related to any physical phenomenon. It is pretty much a "black box."

Schnelle (1990) reports a successful use of a neural network to predict compositions in a distillation system. The neural network was used to predict composition on a minute-by-minute basis to supplement lab results whose frequency was every four hours.

In another industrial example, an impurity in an intermediate chemical was controlled by adjusting a temperature in the process. The impurity in the final product was analyzed in the lab. There was a 12-hour time delay between when the control action occurred and when the analysis was reported on the final product, making it extremely difficult to do control. A six-input, one-output neural

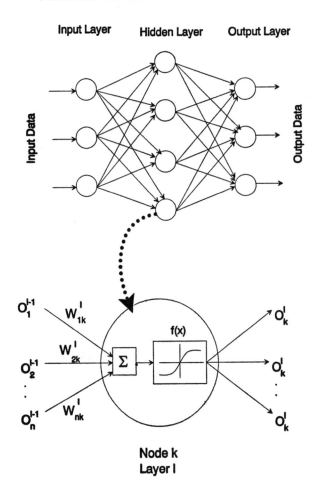

Figure 8–1 A Neural Network System (adapted from Lee and Park by permission of the American Institute of Chemical Engineers)

network was installed that predicted the impurity in the product from a set of process conditions every hour. The predicted impurity became the process variable to an advisory integral-only algorithm that calculated an adjustment to the process temperature. The actual adjustment was performed by the operator. With the neural network advisory controller, the variability was reduced 75% and the Ppk raised from less than 1 to 3.

There appears to be great promise for using these statistical methods for predicting variables that otherwise would only be available infrequently because of time delays in the process and analytical lab.

NEURAL NETWORKS WITH MODEL PREDICTIVE CONTROL

There has also been a lot of interest in using neural network estimators for control where the neural network provides a predicted process variable to the control algorithm. The incentive for using such a scheme is to overcome the shortcomings of a linear model when the process is severely nonlinear. Such situations occur with some polymer reactors, when the process undergoes frequent transitions from one product to another, or when the process operates over a wide dynamic range.

Temeng (1992) reports on an industrial application of a nonlinear model predictive controller that uses a dynamic seven-input, three-output neural network model instead of the conventional linear impulse response model. This *neural model predictive control* was applied to temperature control on a packed bed reactor. They made the normally static estimator dynamic by a time-lag recurrent network [Werbos 1988] in which delayed outputs are fed back to the inputs of the neural network. The controller consisted of an optimizer that calculated the control moves to minimize the sum of squared errors between the setpoints and the projected process variables predicted by the neural network.

Lee (1992) discusses a neural network that provides feedforward action in parallel to an MPC strategy. The neural network learns the patterns between the manipulative variables and the measured disturbances that are not taken into account by the MPC model. In another research paper, Psichogios (1991) applies a neural network in a model-based control strategy where the neural net is both the process model and the controller.

Piovoso (1991) illustrates two uses of a neural network for process control. In the first case, the neural network predicts the process output response to a set of manipulative variables. The predicted outputs are subtracted from the desired response to generate an error. The manipulative variables are iteratively changed to minimize the error. The manipulative variables are then applied to the process. In the second example, the neural network is trained to predict the damping ratio, natural frequency, and gain for a second-order process model. The parameters are then used to guide the selection of PID controller tuning to achieve some criteria of performance.

This new technology is getting a lot of research interest in both academia and industry, and holds promise for dealing with the highly nonlinear systems that pervade the industry. At this point, the controller structures are experimental, but in time, the DCS vendors will no doubt commercialize the ones that appear to offer real value to their customers.

REFERENCES

ARDELL, G. G., and GUMOWSKI, B. Model Prediction for Reactor Control. *Chemical Engineering Progress,* pp. 77–83, June 1983.

DOLLAR, R., et al. Consider Adaptive Multivariable Predictive Controllers. *Hydrocarbon Processing,* pp. 109–112, May 1993.

FRIEDMAN, Y. Z. Avoid Advanced Control Project Mistakes. *Hydrocarbon Processing,* pp. 115–120, October 1992.

HARRIS, T. J., MACGREGOR, J. F., and WRIGHT, J. D. An Overview of Discrete Stochastic Controllers: Generalized PID Algorithms with Dead-Time Compensation. *The Canadian Journal of Chemical Engineering,* Vol. 60, No. 3, pp. 425–432, 1982.

KALMAN, R. E. A New Approach to Linear Filtering and Prediction Problems. *Trans. ASME J. Basic Eng.,* pp. 35–45, March 1960.

KOSANOVICH, K. A., and PIOVOSO, M. J. Process Data Analysis Using Multivariable Statistical Methods. Paper presented at 1991 Automatic Control Conference.

KRESTA, J., MARLIN, T., and MACGREGOR, J. Choosing Inferential Variables Using Projection to Latent Structures with Application to Multicomponent Distillation. Paper 23F, AIChE Annual Meeting, Chicago, November 1990.

LEE, M., and PARK, S. A New Scheme Combining Neural Feedforward Control with ModelPredictive Control. *AICHE Journal,* Vol. 38, No. 2, pp. 193–200, February 1992.

MEJDELL, T., and SKOGESTAD, S. Estimation of Distillation Compositions from Multiple Temperature Measurements Using Partial Least Squares Regression. *Ind. Eng. Chem. Res.* Vol. 30, pp. 2543–2564, 1991.

PIOVOSO, M. J., OWENS, A. J., GUEZ, A., and NILSSEN, E. Neural Network Process Control. Paper presented at ANNA Conference, 1991.

PIOVOSO, M. J., KOSANOVICH, K. A., and YUK, J. P. Process Data Chemometrics. *IEEE Transactions on Instruments and Measurements,* Vol. 41, No. 2, April 1992.

PIOVOSO, M. J., and KOSANOVICH, K. A. Applications of Multivariate Statistical Methods to Process Monitoring and Controller Design. Paper submitted to *International Journal of Control,* June 1993.

PSICHOGIOS, D. C., and UNGAR, L. H. Direct and Indirect Model Based Control Using Artificial Neural Networks. *Ind. Eng. Chem. Res.,* Vol. 30, No. 12, pp. 2564–2573, 1991.

RAY, W. H. *Advanced Process Control.* New York: Mc-Graw-Hill, 1981.

SCHNELLE, P. D., and FLETCHER, J. A. Using Neural Based Process Modeling for Measurement Inference. Paper 90–915 presented at ISA/90 International Conference and Exhibit, New Orleans, October 1990.

TEMENG, K. O., SCHNELLE, P. D., SU, H., and MCAVOY, T. J. Neural Model Predictive Control of an Industrial Packed Bed Reactor. Paper presented at AIChE Annual Meeting, Miami Beach, November 1992.

WELLS, C. H. Application of Modern Estimation and Identification Techniques to Chemical Processes. *AIChE Journal,* Vol. 17, No. 4, pp. 966–973, July 1971.

WERBOS, P. J. Backpropagation through Time: What It Does and How to Do It. *Proc of IEEE,* Vol. 78, No. 10, p. 1550, 1990.

9

Sustaining
the Benefits

It is one thing to design and implement control improvements that gain economic benefits right away, and quite another to sustain the benefits over the long haul. Typically, control benefits deteriorate over time because the systems are not supported [Gerdeman 1990]. If a new control strategy starts to act up, the operator may go to manual control, and the control problem may never be identified and corrected. The result will be a reduction in operating performance in terms of lower yield, higher energy consumption, poorer quality, and less throughput. Unfortunately, no control system or strategy has ever been devised that is completely independent of human support. Thus, sustaining the benefits from control improvements may be just as critical as implementing the improvements in the first place. This fact of life has not been universally accepted. Thus, there are still tremendous opportunities to improve manufacturing in this arena.

CONTINUOUS IMPROVEMENT

Continuous improvement is recommended as the key strategy for sustaining benefits. If you are not improving, odds are you are losing ground. Continuous improvement is the process of making steady, incremental improvements rather than seeking step changes or breakthroughs. Dertouzos (1989) states that world class plants need to continuously improve the quality and reliability of products and processes to remain competitive. It follows that continuous improvement is really key to sustaining benefits from process control.

The things discussed thus far in this book are really parts of the continuous improvement process:

- Analyzing current performance
- Identifying and assessing potential improvements
- Implementing the improvements

However, this is only one cycle in the process of continuous improvement. The cycle must be constantly repeated at some reasonable frequency. Obviously, continuous improvement cannot be another program of the month, but must become part of the manufacturing culture. It may take a lot of work to make it happen in a manufacturing organization. Continuous improvement challenges some long-standing paradigms that say there is no glory in supporting and improving what is already in place. In some cases, the reward system must be changed to promote and encourage continuous improvement in existing plants.

There are several elements that need to be in place to make continuous improvement possible:

- Plant infrastructure
- Training
- Metrics for monitoring performance
- Auditing the results of improvements

INFRASTRUCTURE

Continuous improvement requires that a proper plant infrastructure be in place. Infrastructure has to do with the *permanent foundation* to carry out and sustain the process of continuous improvement. The elements of infrastructure are:

- Supportive management
- Skilled resources

- Tools
- Empowered organization

The management must first of all recognize and believe that improving manufacturing performance is key to a successful business. Furthermore, they must believe that process control is an important part of achieving success in manufacturing and put their support behind it. Management and operators must feel ownership for the process measurement and control systems they operate. The business needs, the required performance from the controls, and the potential value of process control to the business must be clearly understood by all. Without these things, the motivation to improve process control will not be there. Management has the prime responsibility to see that this happens and to empower the organization to make it happen.

The organization must have the resources in terms of skilled people and tools to make and sustain the improvements. There has to be integration of the resources who design and implement the improvements and the people who operate the equipment. It is not acceptable for the designers to just "throw the improvements over the wall" to the operators and expect the systems to be accepted and operated at their designed potential. Operators must be able to rely on technical support when the controls do not perform in a satisfactory manner. Otherwise, they will find some other way to control the process. Engineers must be willing to troubleshoot and fix problems, keep drawings and other documentation up to date, and spend the necessary time to demonstrate and convince the operators that the controls are adding value.

Friedman (1992) suggests that operators need to have a way to document operating problems. Then engineers should document what their diagnosis of the problem was, what was done to fix it, and the results. That way, a history can be collected around equipment to make it easier for future personnel to learn the operation and solve future problems. Of course, institutionalizing any documentation is difficult when so many other more critical needs must be met. It takes a disciplined organization to document anything, especially something with long-term benefits.

One of the facts of life facing many companies now is a lack of adequate technical resources. If this is the case, there are other ways to provide the coverage. Sometimes it is feasible to leverage technical skills over several plant sites through networking. Networks are really teams who share their technical expertise over several sites, and develop standard practices and tools that can be applied across similar businesses. It may also be possible to contract the needed coverage with outside firms or central engineering groups.

All operating and technical personnel must also be familiar with the business needs of the operation, and focus their operating procedures and ongoing improvement efforts on meeting the business objectives. Communication and co-ordination can have dramatic benefits. For example, if the process has several steps that may be operated from different locations and by different crews, com-munication among the operating groups on what each is doing and trying to achieve is critical to make the total process behave as a single well-integrated business unit. In one situation, the lack of communication between process areas at a site resulted in needless downtime in one area simply because maintenance outages were not coordinated.

It is likewise important that the maintenance organization be integrated into the manufacturing activity in such a way that they understand the business needs and feel the urgency of keeping the systems operating properly. Proper mainte-nance is necessary for keeping instruments and controls operating and calibrated properly, which is a very critical element in having a smooth-running, accurate, and reliable manufacturing system. A well-designed preventive maintenance pro-gram is essential to make sure instruments and controls stay on-line and are prop-erly calibrated.

TRAINING

One of the ways to integrate the various pieces of the organization so that they have the functioning capability to perform their respective tasks is through train-ing. Dertouzos (1989) says that on-the-job training is essential because U.S. schools are turning out too few technically and scientifically trained people. Training can be very effective when it is a joint effort between the engineers who design and support the control systems and the operators who use the systems. Each group has something to teach the other.

Process control engineers obviously need to keep up with the state-of-the-art developments in the instrumentation and control technology. They also need a balance of knowledge in both the process control technology and process engi-neering. It is important that they know how equipment operates, key design and operating parameters, and generally accepted control strategies. Sometimes oper-ating problems are caused by equipment malfunctions, so they should understand

and know how to diagnose problems having to do with heat and mass transfer like fouling, foaming, contamination, and so forth.

There are a number of ways training can be obtained if not through one's own organization. Shunta (1989) reports on a training program to raise the skill level of instrument and process control engineers that was developed jointly with and contracted to the Instrument Society of America. This came out of a belief of management that there is enormous economic potential in manufacturing through the appropriate application of process control technology. Process control specialists should be able to design improvements every year that will achieve on average $300,000/year in savings. Engineers were given an opportunity to take as much as eight weeks of training covering basic measurement instrumentation, final control elements, process analyzers, programmable logic controls, single-loop and DCS controls, modeling and simulation, and applications of control strategies to a variety of unit operations. The training has been given regularly twice a year since 1985. In addition to formal training courses, engineers need on-the-job training on how the operators run the process and, in particular, how they respond to abnormal conditions. Knowing how the operator does the job helps the engineer to design improvements that are relevant and appropriate.

Operators need training on control upgrades and improvements, including demonstrations on the value of the improvements, so they can feel ownership for them [Hanley 1990]. Operator training on the fundamentals of process dynamics is also helpful for them to see how process deadtime and other kinds of process behavior affects controllability. Computer simulations are a valuable tool in this type of training, especially when the simulator is connected to the DCS console so that the operators get a feel for the process.

Training for managers should demonstrate the value of the control systems in terms of meeting business objectives. Otherwise, they will tend not to provide the leadership necessary to sustain the benefits. Managers also need to understand how technology and people working together leads to business success.

These may all seem pretty obvious needs for training. Rather than stop here, however, it is recommended that a manufacturing organization take the time to discover their most pressing training needs, and then develop a personnel training and development program to address them. You may discover a need that is unique to your organization and culture. For example, it may be that the biggest problems in your operation stem from mistuned control loops. Thus, you may want to assess your whole activity around servicing control loops by asking questions such as: Who is responsible? Do they have the skills and time? Is the training adequate? Some other organization may have a critical need around analyzer maintenance, and so on.

The critical question is not so much how to get the training but what and how much training is needed. If in-house experts cannot provide the training, there are many other options available through contractors, vendors, schools, and professional societies.

METRICS FOR MONITORING PERFORMANCE

A critical element in the continuous improvement cycle is the use of metrics to gauge performance. Metrics not only show current performance, but also compare performance against the desired goal. One company has a standard practice for how operating metrics will be applied. The recommendations have been implemented in the DCS and have received great acceptance from the operating personnel.

Two types of control system performance metrics are covered in this chapter: *utilization metrics* and *variability metrics*. Utilization metrics measure how much of the time critical measurements and controls are fully employed. Gerdeman (1990) reports on a utilization program that monitors the percentage of utilization of an advanced control system and the corresponding savings gained. The program clearly illustrates that savings increase as the percentage of utilization increases. Utilization is a direct result of the support provided by an advanced control support group. The report also shows the potential savings for each control loop at 100% utilization. This helps to prioritize which loops need to be serviced. Gerdeman (1990) also shows that regulatory controls in the DCS experience a relatively high utilization. However, advanced computer loops that rely on analytical measurements have the lowest utility because of low uptime of the analyzers.

It is also important to keep a record of the frequency and duration of alarm and interlock violations. Since most, if not all, DCSs provide this capability, we will only mention it here.

Metrics for Critical Measurements and Controls

A number of utilization metrics are suggested below. The metrics can be implemented in a number of ways, but the important thing is to review and act on the metrics in a systematic way; in other words, institutionalize their use.

Utility of analyzers and corresponding composition control loops

$$\% \text{ Utility of Composition Control Loop } = \frac{\text{Time Control Loop in Cascade}}{\text{Time Process is Running}} \tag{9-1}$$

$$\% \text{ Availability of Analyzer} = \frac{\text{Time Analyzer is Operating and Calibrated}}{\text{Time Process is Running}} \tag{9-2}$$

When a predictive model is used to supplement or replace the analyzer, it should be monitored in the same way because of the importance of its results.

Utility of computer control loops

Computer loops are those that contain custom algorithms and code for special applications beyond the regulatory control functions found in the base-level DCS configurations. These loops are usually associated with critical controlled variables.

$$\% \text{ Utility of Computer Control Loop } = \frac{\text{Time Control Loop in Computer Mode}}{\text{Time Process is Running}} \tag{9-3}$$

$$\% \text{ Availability of Computer } = \frac{\text{Time Computer is On-Line}}{\text{Time Process is Running}} \tag{9-4}$$

Utility of regulatory control loops (e.g., DCS)

$$\% \text{Utility of Control Loop } = \frac{\text{Time Loop in Cascade or Automatic}}{\text{Time Process is Running}} \tag{9-5}$$

Variability Metrics

Another valuable metric measures the variability of key product properties and process variables. Kane (1986) suggests that process capability indices be used to monitor the spread of the data relative to the specifications (Cp) and how far the average is from the target or aim (Cpk). If Cpk equals Cp, the process is on-aim. If Cp is less than the goal, the common-cause variability is larger than desired, in which case we should identify and eliminate the root causes or change the process (see Chapters 2 and 3).

We can also monitor the performance index Pp to see if the special-cause variability is greater than desired (which would indicate a problem with the controls) and Ppk to see if the process average is off-aim. For product properties we would use the customer specifications in the calculation, and for process variables we would use the corresponding standard operating conditions or ranges (SOCs).

A simple metric to compare variability against a benchmark for ideal feedback control is given by Equation 9–6.

$$\text{Variability Index} = \frac{S_{\text{tot}}}{S_{\text{apc}}} \qquad (9\text{–}6)$$

where S_{apc} is the minimum standard deviation we can expect with feedback control operating with the current measurement, and S_{tot} is the total standard deviation. A variability index of 1.0 is the goal. The index increases with poorer control.

Another index we could use includes the deviation from aim along with variability.

$$\text{Control Index} = \frac{\sqrt{S_{\text{tot}}^2 + (\text{aim} - \text{average})^2}}{S_{\text{apc}}} \qquad (9\text{–}7)$$

The control index goal is 1.0. When used with Equation 9–6, this index shows if the process is off-aim. The aim value will typically be the controller setpoint for a process variable. Other benchmark parameters instead of S_{apc} can be applied when appropriate. Note from Chapter 2 that in the calculation of S_{apc}, S_{cap} must be limited to less than $\sqrt{2}\,S_{\text{tot}}$ to avoid taking the square root of a negative number. In theory, S_{cap} should be less than $\sqrt{2}\,S_{\text{tot}}$, but this could be violated in a finite amount of data. If $S_{\text{cap}} > \sqrt{2}\,S_{\text{tot}}$, S_{apc} should be considered zero.

Equation 9–7 can be easily modified to estimate the potential percentage of reduction in standard deviation or *improvement* for control in Equation 9–8.

$$\text{Percentage of Improvement} = 100 \left(1 - \frac{1}{\text{Control Index}} \right) \qquad (9\text{–}8)$$

RELATING VARIABILITY TO BUSINESS GOALS

It is helpful to relate variability to business metrics so that all personnel can appreciate the benefits of keeping the controls operating effectively. Chapter 2 shows how the percentage out-of-specifications can be calculated for a normally distributed variable given the total standard deviation, the specifications or SOCs, and the average value. We can apply the same equations to estimate the percentage of product that meets specification limits or, alternatively, the percentage of the time a process variable falls within the SOC limits. We will refer to the state of being within acceptable limits as *conformance*. In some cases, conformance translates directly to first-pass first-quality yield. This would be true if the product not conforming to the specification limits had to be reworked, blended, or sold at a lower price. Process variables conforming to limits in other cases may relate to throughput or minimum energy consumption.

Conformance is estimated as follows. First calculate Z_L and Z_H from Equations 9–9 and 9–10 for the product property or process variable X.

$$Z_L = \frac{\overline{X} - X_L}{S_{tot}} \tag{9-9}$$

$$Z_H = \frac{X_H - \overline{X}}{S_{tot}} \tag{9-10}$$

where

$\overline{X}$ = average value
X_H, X_L = high and low limits

Second, look up the values for Z_L and Z_H in Table 9–1 and find the corresponding conformance. The overall conformance percent is obtained by adding the two conformances and subtracting 100.

ESTIMATING STANDARD DEVIATION ON-LINE

In order to use on-line performance metrics like Equations 9–6, 9–7, and 9–8, we need to consider how variability is estimated. One way is to periodically collect a set of historical data and perform the traditional batch calculations of variance

TABLE 9–1

Negative Z-Values		Positive Z-Values	
Z-Value	Conformance	Z-Value	Conformance
–0.0	50.000	+0.0	50.00
–0.1	46.017	+0.1	53.98
–0.2	42.074	+0.2	57.93
–0.3	38.209	+0.3	61.79
–0.4	34.458	+0.4	65.54
–0.5	30.854	+0.5	69.15
–0.6	27.425	+0.6	72.57
–0.7	24.196	+0.7	75.80
–0.8	21.186	+0.8	78.81
–0.9	18.406	+0.9	81.59
–1.0	15.866	+1.0	84.13
–1.1	13.567	+1.1	86.43
–1.2	11.507	+1.2	88.49
–1.3	9.680	+1.3	90.32
–1.4	8.076	+1.4	91.92
–1.5	6.681	+1.5	93.32
–1.6	5.480	+1.6	94.52
–1.7	4.457	+1.7	95.54
–1.8	3.593	+1.8	96.41
–1.9	2.872	+1.9	97.13
–2.0	2.275	+2.0	97.72
–2.1	1.786	+2.1	98.21
–2.2	1.390	+2.2	98.61
–2.3	1.072	+2.3	98.93
–2.4	0.820	+2.4	99.18
–2.5	0.621	+2.5	99.38
–2.6	0.466	+2.6	99.53
–2.7	0.347	+2.7	99.65
–2.8	0.256	+2.8	99.74
–2.9	0.187	+2.9	99.81
–3.0	0.135	+3.0	99.86
–3.1	0.097	+3.1	99.90
–3.2	0.069	+3.2	99.93
–3.3	0.048	+3.3	99.95
–3.4	0.034	+3.4	99.97
–3.5	0.023	+3.5	99.98
–3.6	0.016	+3.6	99.98
–3.7	0.011	+3.7	99.99
–3.8	0.007	+3.8	99.99
–3.9	0.005	+3.9	100.00

and standard deviation as shown in Chapter 2. However, this method will not provide minute-by-minute monitoring. There are two approaches for estimating variability on a current basis. One is called the *moving window* method, where the calculations are performed on a fixed-length set of data. The data set is updated with a new data point when it becomes available, and the oldest data point is then discarded. The calculations are performed every time the data set is updated. An obvious question is, how large should the data set or window be? That will be discussed later.

A second approach is to estimate variability by what control engineers call a smoothing filter. Statisticians might call it an *exponentially weighted moving estimator*. The filter assigns a weight to each new data point and adds it to a weighted value of the previous estimate. The sum is then the new estimate. To illustrate, suppose we want to estimate the current variance V_j using the filter. The equation is expressed below.

$$V_j = r(X_j - \overline{X})^2 + (1-r)\ V_{j-1} \tag{9-11}$$

where

X_j = current process variable or product property

$\overline{X}$ = average value of X

$X_j - \overline{X}$ = error e

r = weighting factor $(0-1)$

The coefficients on the right-hand terms in the equation sum to one. The weighting factor determines the rate at which past estimates of variance are discounted. Therefore, the weighting factor can be adjusted to estimate long-term variability (r is small) or short-term variability (r is large). MacGregor (1993) refers to this estimator as the *exponentially weighted moving variance* (EWMV).

Recall that the standard deviation S is the square root of variance. Equation 9–12 gives an estimate of S called the *exponentially weighted moving standard deviation* (EWMSD).

$$S_j = \sqrt{r(X_j - X_{sp})^2 + (1-r)\ S_{j-1}^2} \tag{9-12}$$

Equation 9–12 can be used to estimate S_{tot} in Equations 9–6 and 9–7.

For this we want an estimate of the long-term standard deviation so r will be small to give more weight to the past values. For estimating short-term variability, we use a larger value of r to discount the past variability more quickly and give more weight to the current process value.

$\overline{X}$ in Equations 9–11 and 9–12 may or may not be constant. If $\overline{X}$ varies, as it probably will in a chemical process, we can estimate it using an *exponentially weighted moving average* (EWMA), as shown in Equation 9–13.

$$\overline{X}_j = k\,X_j + (1-k)\,X_{j-1} \tag{9–13}$$

Again, k is a weighting factor that determines how fast the previous estimate of $\overline{X}$ is discounted. A different symbol is used for weighting factor because sometimes it may be appropriate to use different values. Using a small value of k in Equation 9–13 better estimates the long-term average. The weighting really depends on how the metric will be used.

By using the EWMA estimate of $\overline{X}$ in Equation 9–12, the estimate of standard deviation will not pick up shifts in the average as will a fixed value of $\overline{X}$.

If the process variable is being controlled, we might like to have a metric that picks up when the process has drifted away from the setpoint or aim due to a process upset. We can achieve this by substituting the controller setpoint X_{sp} for $\overline{X}$ in Equation 9–12.

$$S_j = \sqrt{r(X_j - X_{sp})^2 + (1-r)\,S_{j-1}^2} \tag{9–14}$$

MacGregor (1993) refers to this metric as the *exponentially weighted root mean square* (EWRMS). Equation 9–14 also estimates the numerator of the control index in Equation 9–7.

We can recursively estimate the capability standard deviation. Equation 9–15 estimates the capability standard deviation by the *exponentially weighted moving MSSD standard deviation* (EWMMSD).

$$S_{\mathrm{cap}(j)} = \sqrt{r\,\frac{(X_j - X_{(j-1)})^2}{2} + (1-r)\,S_{\mathrm{cap}(j-1)}^2} \tag{9–15}$$

We can then estimate the minimum variance possible with automatic feedback control on line from Equation 9–16.

$$S_{\mathrm{apc}(j)} = S_{\mathrm{cap}(j)}\sqrt{2 - \left[\frac{S_{\mathrm{cap}(j)}}{S_{\mathrm{tot}(j)}}\right]^2} \tag{9–16}$$

WEIGHTING FACTOR

Setting the weighting factor may seem more of an art than a science. It might be of some help to relate it to something having more physical significance. Since the estimator is really a first-order filter we can relate r or k to the filter time constant T. The differential equation for a first-order filter operating on variance V is

$$\frac{dV}{dt} + \left(\frac{1}{T}\right) V = \left(\frac{1}{T}\right) e^2 \tag{9-17}$$

where T is the *time constant* or time it takes to reach 63% of the steady-state value following a step change in e $(X - \overline{X})$. The discrete version of Equation 9–17 is given by Equation 9–18.

$$\frac{(V_j - V_{j-1})}{Ts} + \left(\frac{1}{T}\right) V_j = \left(\frac{1}{T}\right) e_j^2 \tag{9-18}$$

where Ts is the sampling period in minutes.

We can now solve for V_j by Equation 9–19.

$$V_j = \left(\frac{Ts}{T + Ts}\right) e_j^2 + \left(\frac{T}{T + Ts}\right) V_{j-1} \tag{9-19}$$

where

$$r = \left(\frac{Ts}{T + Ts}\right) \quad \text{and} \quad 1 - r = \left(\frac{T}{T + Ts}\right)$$

Since we know the sampling period for the data, we can relate r to the time constant of the response. For some this may be an easier way to visualize or tune the filter weighting factor parameter.

APPLICATION OF ESTIMATORS

The on-line estimating techniques described above were applied to a typical process modeled as a first-order lag plus deadtime having a period of oscillation of five minutes (see Figure 7–1). The process variable, concentration, was sampled and controlled every minute. In the first example, the process was subjected to a normal random disturbance in flow A to the process. The time response is

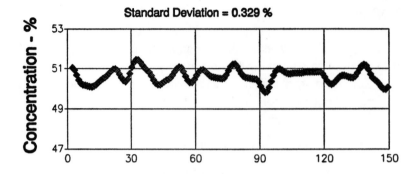

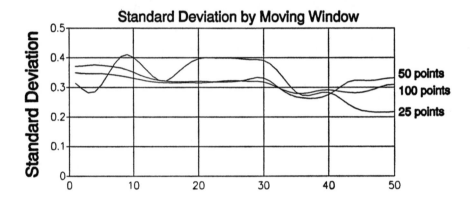

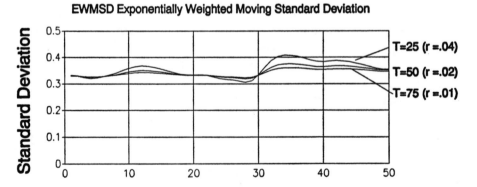

Figure 9–1 Comparison of Moving Window and EWMSD for Estimating Standard Deviation

shown in the top graph of Figure 9–1. The total standard deviation, measured over 150 minutes of data, was 0.329%.

Comparing the Moving Window and EWMSD Estimators

The moving window and the EWMSD (Equation 9–12) are compared for estimating the standard deviation. The performance of the estimators is shown in Figure 9–1. Three moving window estimators having sizes of 25, 50, and 100 data points are shown in the center graph. The window having 25 points had quite a bit of variability, but doubling its size to 50 points provided a much better long-term estimate of total standard deviation S_{tot}. Fifty data points describes about 10 periods of oscillation. Making the window larger did not have a significant effect.

The bottom graph shows the EWMSD estimator for three weighting factors r. The weighting is also stated in terms of the filter time constant T. The 25-minute time constant estimator was the most oscillatory. There was not much difference between $T = 50$ and $T = 75$ minutes. This indicates that a filter time constant of 50 minutes, which covers about 10 periods of oscillation, seems to estimate S_{tot} fairly well. Judging from these examples, the EWMSD compares favorably to a moving window estimator. It also has an advantage in needing less computer memory.

MSSD (Capability) Standard Deviation Estimator

Figure 9–2 shows an example of estimating the MSSD standard deviation on-line using Equation 9–15. The three values of weighting factors are repeated. Again, the weighting factor corresponding to a filter time constant of 50 minutes seems adequate for estimating capability standard deviation.

EWMA Estimation of Process Average

The next example demonstrates the EWMA (Equation 9–13) for estimating the average value of the process variable following a setpoint change in the concentration controller. Figure 9–3 compares three estimators having weighting factors k of 0.02, 0.05, and 0.1 to the actual data. The larger the k value, the more closely

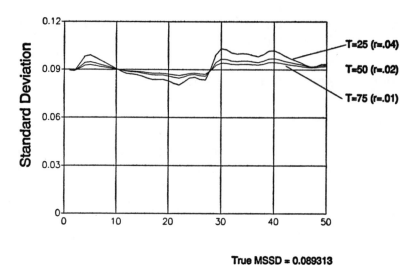

True MSSD = 0.089313

Figure 9–2   Exponentially Weighted Moving MSSD Standard Deviation
(EWMMSD)

the estimator follows the data. The small values are better for making long-term estimates of the process average.

Load Disturbance

In the next example, the process is subjected to a 5% step change in flow A. EWMSD and EWMMSD are applied to estimate the total standard deviation and capability standard deviation, respectively. The controller setpoint was substituted into the EWMSD equation for the process average so that it would reflect variability due to load upsets but not setpoint changes. The weighting factor r was specified to be 0.05 in both estimators so that new data was discounted 95%. The two metrics were used to calculate the minimum expected standard deviation with good feedback control S_{apc} from Equation 9–16. A feed rate change in flow A was introduced 20 minutes into the run. The estimates are shown in Figure 9–4. The concentration did drop after the upset, but the concentration controller quickly got the concentration back to setpoint, so the standard deviation showed little increase. The difference between S_{tot} and S_{apc} is the estimated potential improvement in standard deviation for improved control, and it is only about 0.2%.

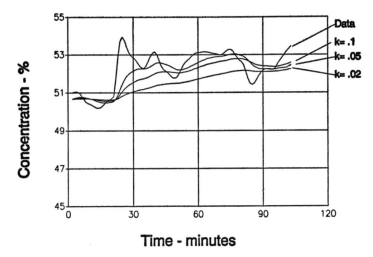

Time - minutes

Figure 9–3　Exponentially Weighted Moving Average (EWMA)

Change in Variability

In the final example, the variability of the load upset was changed. The standard deviation of flow A was changed from 25% to 50% of maximum flow 20 minutes into the run. This is shown in Figure 9–5. The change is readily apparent by the standard deviation curves but is not so pronounced in the operating data. This example shows the value in having information like this on key product properties and process variables to signal to the operator which loops are not performing as they should.

It would also be useful to have a plot of the specification limits or SOCs, the average value, and the total standard deviation—a sort of dynamic histogram like Figure 9–6. These are all possible from the on-line estimators. Of course, this is exactly the purpose of control charts like Shewhart charts. Certainly, there are numerous metrics that can be applied, depending on the preferences of the operating people. The important thing is to *use metrics to monitor performance and keep the measurements and control loops gaining the economic benefits*. It is also recommended that savings be reported along with the performance metrics to keep everyone focused on achieving and sustaining the benefits.

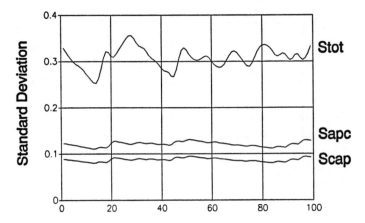

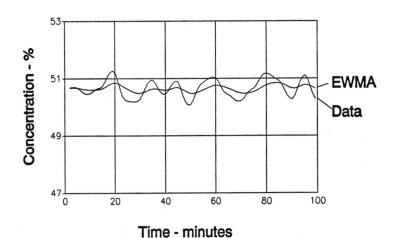

Time - minutes

Figure 9–4 Exponentially Weighted Moving Estimate of Standard Deviation
for Load Disturbance

SPC MONITORING CHARTS

The classic way in which statisticians have approached quality control has been
through the use of control charts that monitor quality parameters and determine
when the process is not in statistical process control. The use of control charts is
also an excellent way to continuously improve the process by finding special

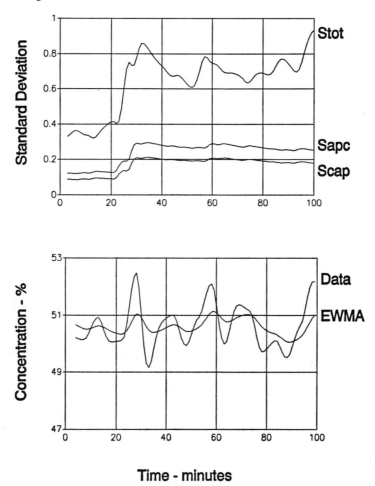

Figure 9–5 Exponentially Weighted Moving Estimate of Standard Deviation
for Variability Change

causes of variation as they come up and then eliminating them. Control charts are
very much akin to performance metrics. This section discusses how control
charts are used to monitor the process. In the next section, SPC charts are used in
conjunction with APC to optimize process control.

The earliest and simplest quality control chart was introduced by Shewhart
in 1931. This is also the one most commonly used in industry. An example is
shown in Figure 9–7. The product property or process variable being monitored
is plotted chronologically. Each data point can represent a single value or the av-

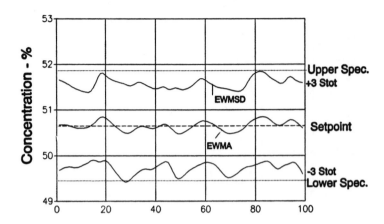

Figure 9-6   An On-Line Histogram for Monitoring the Process

erage of several data points. The vertical axis shows the engineering units, and the center is the aim or setpoint near which the variable must be controlled. The chart also shows upper and lower control limits set at ± 3 standard deviations from the aim. The standard deviation is in fact the capability standard deviation S_{cap}. If the process is in the state of statistical process control, the data will fall

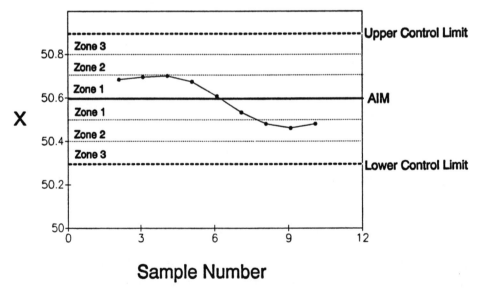

Figure 9-7   A Shewhart Control Chart

almost entirely within the control limits. An alarm occurs when a data point falls beyond the 3 standard deviation limit, signifying that the process is no longer in a state of statistical process control and a special cause has likely occurred. The basis for using S_{cap} is that the process is normally expected to be in the state of statistical process control.

The standard Shewhart alarm rule is not always optimal, so there are a number of other rules that can be applied to detect when a special cause has occurred. One or more of these rules can be applied to fit a particular situation. Some examples are shown in Table 9–2. Since these rules may call for action before the data reaches the 3 standard deviation limit, the capability standard deviation may be too tight. Therefore, a different standard deviation may be recommended.

One of the disadvantages of the Shewhart chart is that it is relatively insensitive to persistent, moderate changes and overly sensitive to extreme values or outliers. For that reason, CUSUM control charts have been used predominately in some companies [Lucas 1976]. CUSUM, as the name implies, plots the cumulative sum of deviations from the target.

$$\sum_{i=1}^{n}(X_i - \text{target})$$

It has the advantage of being sensitive to moderate changes without being overly complicated. Thus, it is well suited to detecting small shifts, which characterize many applications in the CPI.

A not so widely used statistical monitoring technique, but one that appeals particularly to control engineers, is the EWMA control chart, which plots the product property or process variable values calculated by Equation 9–13. The tuning parameter k, which determines how quickly past values are discounted, is a tuning parameter. As k decreases, past values are given more weight, and the EWMA approaches the CUSUM method. As k increases to 1.0, only the current point is weighted, and the EWMA chart will be equivalent to the Shewhart chart.

TABLE 9–2   SHEWHART CONTROL CHART RULES

A. One or more points outside control limits
B. Two out of three successive points on the same side of the average in Zone 3 or beyond
C. Four out of five successive points on the same side of the average in Zone 2 or beyond
D. Nine successive points on one side of the average

A common value for k is 0.2. The EWMA has one rule: a signal is given when the EWMA goes outside of limits [Kittlitz 1987, Hunter 1986], where the limits are

$$\text{Limits} = \pm 3 \, S_{\text{cap}} \sqrt{\frac{k}{2-k}}$$

The advantage of the EWMA is that the average is insensitive to extreme values. It is also conceptually simple and easy to implement in a DCS. A recent paper by Lucas (1990) goes into some detail about the best way to implement the EWMA monitor.

Thus, through the use of SPC charts, we have a mechanism for determining when it is appropriate to seek out a special cause of variability. These charts can be used in conjunction with automatic process control (APC) in a combined control strategy, as shown in the next section.

COMBINING APC AND SPC

A number of papers have appeared in the last few years in which SPC and APC have been integrated for optimal quality control. VanderWiel (1992) introduced what he and his colleagues have named algorithmic statistical process control (ASPC). The control loop contains an APC algorithm for frequently adjusting the manipulative variable to reduce variability and SPC algorithms for monitoring. The SPC algorithms (Shewhart, CUSUM, or EWMA charts) monitor the process variable to see that APC is doing its job properly, and the APC output (manipulative variable) to determine if it is outside the standard operating range. If it is, the SPC algorithm alerts the operator that it has detected a special cause which should be identified and eliminated if feasible. If this strategy works properly, the process will follow its setpoint with minimum variability, the process variable will not be autocorrelated, and special causes of variability will be eliminated. Thus, ASPC delivers the best of both worlds: the short-term effects of upsets are compensated for, and in the long term, the process is improved. This strategy is illustrated in Figure 9–8. Employing such a strategy is a sound way to ensure continuous improvement.

Deshpande (1993) points out that since SPC algorithms are based on the data being not autocorrelated but independent and normally distributed, other steps need to be taken when the data are correlated (because the control is not minimum variance). Otherwise, the SPC monitor may produce a large number of

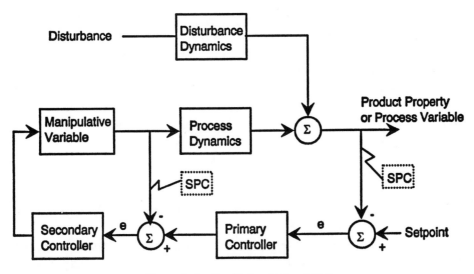

Figure 9-8 Combining APC and SPC

false alarms because of the control limits being set too closely together. Deshpande (1993) suggests two approaches:

- Fit an appropriate time series model to past observations and apply the control charts to the residuals (i.e., the deviations between the data and the predictions). This is a sound approach, but the downside is that time series analysis requires expertise that many general practitioners may not have.
- Set the sampling period long enough so that the autocorrelation coefficient is sufficiently small. Although this will work, it is not very desirable since you lose information about the process between sampling periods.

A third method is to expand the control limits to get an acceptably small false alarm rate. This may require some trial and error, although there has been some work by statisticians to calculate acceptable limits. (Statisticians speak in terms of getting an acceptable average run length [ARL], i.e., the time period between false alarms.) One approach is to identify a period of acceptable variability and calculate the total standard deviation for the data. This standard deviation would then be used in the control limits; or, if we would normally expect the variability to be at the level given by S_{apc}, we could use S_{apc} in the control limits. There are no hard and fast rules, and the standard deviation selected should be applicable to the specific situation and dependent on what the alarm is intended to signify.

In any event, the use of statistical monitoring coupled with APC has been beneficial. VanderWiel (1992) reports a 35% reduction in the standard deviation of viscosity in a polymer reactor and elimination of off-specification material by applying the ASPC strategy. The strategy employed a model-based minimum variance controller with a CUSUM monitoring algorithm to detect special causes.

SEARCHING FOR ROOT CAUSES

The assumption underlying control charting is that when a special cause of variability is detected, an attempt will be made to discover the root cause and eliminate it. This is a crucial part of continuous improvement in manufacturing. It ought to be an integral part of the job of manufacturing and maintenance personnel. Sometimes the automatic control system can compensate for the upset. However, eliminating the source of the variability instead of moving it somewhere else by automatic controls makes a lot of sense. On the other hand, if the control system cannot compensate for the upset, eliminating it is the only solution unless you are willing to live with the problem. Actually, living with problems happens too often. We tend to accept some process problems as if they are part of the process's personality just as we sometimes learn to live with quirks in people.

Not all special causes will be easily identified, and sometimes an investigation may be in order. The statisticians have a tool that can be applied to aid in the discovery of root causes: the cause-and-effect diagram. This is a simple graphic tool that helps to systematically identify and display the possible causes of symptoms. The cause-and-effect diagram is also called the Ishikawa diagram, named after Professor Kaoru Ishikawa of the University of Tokyo who developed the diagram in 1953. It is also called the fishbone diagram because it resembles the skeleton of a fish. An example is given in Figure 9–9.

Typically, the tool is used to facilitate a brain-storming session with the process experts. The first thing to do is to state the symptom or problem, which becomes the backbone, if you will, of the skeleton. Then write down the major categories of the possible causes; for example, manpower, materials, process equipment, procedures, measurements, and controls. These are the main bones connected to the backbone of the diagram. Then start developing some underlying causes in each of the main categories and add them to the diagram as branches of the main categories. Keep digging by asking the question "why?"

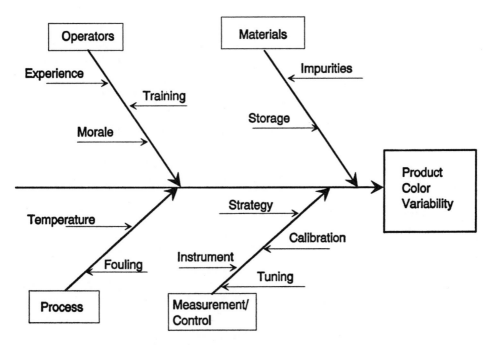

Figure 9–9   Cause-and-Effect Diagram

several times until you think you have exhausted all the possible causes. Select the causes that are most probable and target them for further study to find out if they are really the root cause. That study will involve collecting data to prove or disprove that each is the root cause.

USE OF EXPERT SYSTEMS FOR TROUBLESHOOTING

The above approach utilizes the best thinking of the process experts working together in a brainstorming fashion. Another approach is to use expert system software to capture and mechanize the thinking of the experts. Expert systems have caught on in a big way for diagnosing problems and troubleshooting. If the possible causes can be logically described before the fact, this is a good method because it saves time and effort.

One example of an expert system for troubleshooting a piece of equipment is reported by Rowan (1987). This expert system was developed to troubleshoot the operation of a distillation column. The system runs off-line in a commercial expert system *shell*. There are various commercial programs available today that run on either a PC or minicomputer. Rowan's program is quite typical in that it requires the engineer to answer a series of questions that a consultant might ask. The CRT presents the questions, along with possible true/false or multiple-choice answers, which are then selected. The questions are designed to identify the root causes of the problem. Questions are asked regarding things like the magnitude of the key process variables, condition of the equipment, controller tuning, magnitude of various upsets, and the presence of abnormal conditions such as fouling, flooding, and so on. Based upon the answers, the predesigned logic arrives at a potential solution(s). In some systems, probabilities are assigned to possible solutions instead of just presenting one solution.

Figure 9–10 shows a part of the logic in Rowan's expert system that deals specifically with operating conditions. This is just one part of the overall logical scheme. The logic first checks to see that the operating conditions are normal. If not, each process variable (vapor boilup, reflux flow) is checked for out-of-limit conditions. If a variable is high, the system asks the engineer to reduce the rate and see if the column stabilizes. The answers to these and similar questions are used later to determine the root cause of the problem.

Expert systems have been found to have many uses besides the one just cited. For example, they have been used to:

- Diagnose equipment malfunctions
- Determine the time for preventative maintenance by detecting the presence of fouling, corrosion, and so on
- Determine the time for a process upgrade like changing the catalyst, filter media, and so on
- Automatically tune controllers
- Monitor and guide the startup and shutdown of equipment
- Select appropriate control strategies

The value of expert systems is that they supplement scarce resources and save time in maintaining continuous operation, which leads to economic benefits. On the other hand, this value does not come free. Expert systems represent additional software that must be maintained and supported. Hopefully, we will be seeing more and more of this capability becoming part of the DCS software like automatic controller tuning capability is.

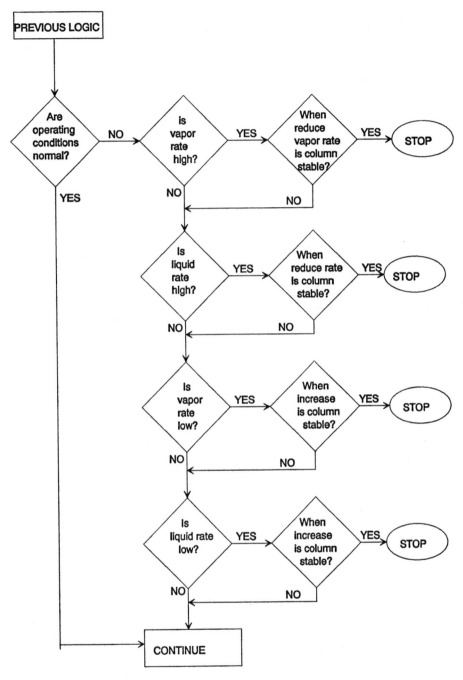

Figure 9–10 An Example of Expert System Logic for Troubleshooting a Distillation Column (Reprinted by permission of W. R. Ellingsen and D. A. Rowan)

IDENTIFYING PROCESS BEHAVIOR BY HISTOGRAMS

It is often difficult to look at a sequence of operating data and determine if there is anything wrong with the process. That is why various statistical metrics are recommended to get a better understanding of what is going on. Sometimes histograms of the data can tell us how the process is operating. A few examples are shown in this section to illustrate. Figures 9–11 and 9–12 show the operating data

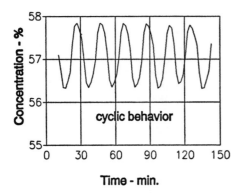
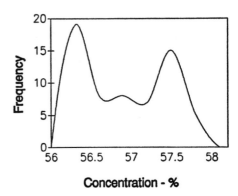

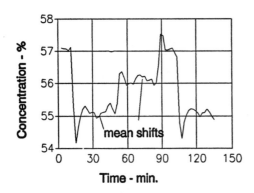
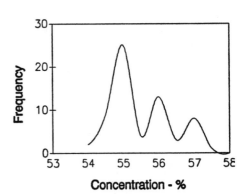

Figure 9–11   A Histogram for Cyclic Behavior and Mean Shifts

and the corresponding histograms for several types of process behavior, namely, cycling, mean or average shifts, good and bad automatic control, and a nonlinear response.

The top part of Figure 9–11 shows cyclic behavior. Cyclic behavior shows up in a histogram as two peaks with a valley in between. The two peaks occur at the two extremes of the cycle where most of the data lie. The lower curves in the

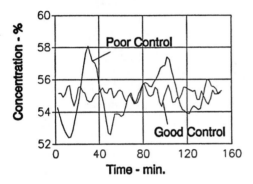
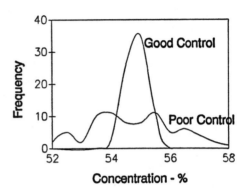

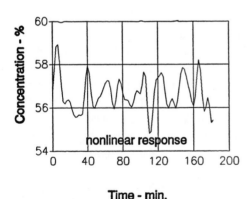
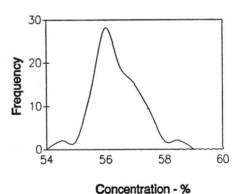

Figure 9–12   A Histogram for Good and Bad Control and Nonlinear Responses

figure illustrate a mean shift. Mean shifts show up as peaks in a histogram. Each new mean has a corresponding peak. The more peaks we see in the data, the more suspect control becomes in terms of good performance. Ideally, we would not expect to see more than one peak, particularly for quality variables.

Figure 9–12 compares the histograms for good and poor control. Good control has a bell-shaped curve, while poor control has a flat and bumpy curve, which reflects wide variability. There are several small peaks that do not correspond to mean shifts, but just the erratic, cyclic behavior of the data. The lower curves in Figure 9–12 show the effect of a process nonlinearity on the histogram. As discussed in Chapter 2, a nonlinearity exhibits skewed distribution. If the skew is pronounced enough, it might prompt some action to try to minimize or eliminate the nonlinearity since it results in poor control.

The problem with using histograms to analyze a process is that the shape is often affected by many things, so they are not as reliable as some of the other metrics presented in this chapter. However, they do tend to distinguish between good and bad operation by the spread and smoothness of the curve. They are just one more tool that can be used to diagnose the health of a control strategy. Fortunately, histograms are already standard features in many statistical software packages.

AUDITING THE RESULTS FROM IMPROVEMENT EFFORTS

This book has stressed how better process control can improve manufacturing by reducing the variability of its products. As important as those initial efforts at continuous improvement are, without constant vigilance, the benefits will probably be lost in time. Periodically, it is prudent to reassess what has been achieved and what can still be gained; in other words, to audit the improvement process. The steps in an auditing process are listed in Figure 9–13.

The first step is to establish what improvement opportunities have been implemented and compare them to the original recommendations to see what remains to be done. Part of this assessment should be to verify the original savings or earnings basis. If the business situation has changed, there may be valid reasons that some of the original recommendations were not implemented. For example, if the original stake was estimated on the basis of a sold-out market and that no longer applies, the opportunity to increase throughput may no longer be attractive. The savings achieved from the implemented opportunities should be quantified and compared to the original savings estimate, and any differences should be reconciled.

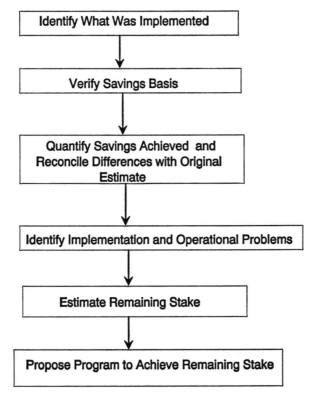

Figure 9–13 Auditing Results from Improvement Efforts

There may have been problems in trying to implement the recommended improvements, and these should be identified. The auditors can then make an assessment as to the feasibility of moving forward on the original recommendations or perhaps suggest an alternate approach.

Finally, the remaining stake that was not captured on the previous improvement cycle should be estimated and plans developed to obtain it. The remaining stake should include any new opportunities that can be identified as well.

Repeating the audit on an ongoing basis, perhaps once a year, ensures that the competitive performance is sustained. This process of self-examination and continuous improvement will provide long-term benefits that otherwise will probably be lost.

REFERENCES

DERTOUZOS, M. L., et al. *Made In America*. Cambridge, MA: MIT Press, 1989.

DESHPANDE, P. B., et al. Achieve Total Quality Control of Continuous Processes. *Chemical Engineering Progress*, pp. 59–66, July 1993.

FRIEDMAN, Y. Z. Avoid Advanced Control Project Mistakes. *Hydrocarbon Processing*, pp. 115–120, October 1992.

GERDEMAN, A. M. *Advanced Control Utilization History and Reporting in a Refinery*. Paper presented at the 1990 ISA International Conference and Exhibit, Philadelphia, October 1990.

HANLEY, J. P. How to Keep Control Loops in Service. *Intech*, Vol. 37, No. 10, October, 1990.

HUNTER, J. S. The Exponentially Weighted Moving Average. *Journal of Quality Technology*, Vol. 18, pp. 203–210, 1986.

KANE, V. E. Process Capability Indices. *Journal of Quality Technology*, Vol. 18, No. 1, January 1986.

KITTLITZ, R., et al. *Quality Assurance for the Chemical and Process Industries*. Amer. Soc. for Qual. Cont., 1987.

LUCAS, J. M. The Design and Use of V-Mask Control Schemes. *Journal of Quality Technology*, Vol. 8, pp. 1–12, 1976.

LUCAS, J. M., and SACCUCCI, M. S. Exponentially Weighted Moving Average Control Schemes: Properties and Enhancements. *Technometrics*, Vol. 32, No. 1, pp. 1–29, February 1990.

MACGREGOR, J. F., and HARRIS, T. J. The Exponentially Weighted Moving Average. *Journal of Quality Technology*, Vol. 25, No. 2, April 1993.

ROWAN, D. A., and ELLINGSEN, W. R. *Expert System for Troubleshooting Distillation Columns*. Paper presented at the 1987 American Society of Engineering Education meeting, Southeastern Massachusetts University, 1987.

SHUNTA, J. P. Process Control Training Yields Big Dividends. *Intech*, p. 74, January 1989.

VANDERWIEL, S. A., et al. Algorithmic Statistical Process Control: Concepts and an Application. *Technometrics*, Vol. 34, No. 3, pp. 286–297, August 1992.

10

Process Design for Improved Controllability

The focus of this book has been on finding ways in which process control can improve manufacturing performance by controlling the variability of product properties and process variables in existing processes. It has been shown that process design affects how well variability can be controlled. The design affects the dynamics of the process, which not only determines how the process responds over time to various upsets but even how stable the process is. The control engineer brings a unique understanding of process dynamics and how the process design can be improved to enhance controllability to meet the business goals.

This chapter focuses on design considerations from a controllability viewpoint and the benefits of making the design compatible with good control.

OPPORTUNITY FOR IMPROVED CONTROLLABILITY THROUGH PROCESS DESIGN

A great opportunity exists to ensure that a process design is compatible with the goals for controlling variability by making sure control aspects are considered in the design of new and modernized plants. The control engineer should be involved in the design at an early enough stage to influence the design before the design documents are finalized and equipment purchased. Typically, this occurs as the preliminary process flow diagrams are being developed, but in some cases it may be as early as product or process development.

The benefits for addressing control in the process design are numerous:

- Getting the design right the first time results in increased capital productivity because fewer dollars are spent fixing problems that can be avoided.
- Capital productivity is again improved by not overdesigning to compensate for lack of understanding of how the process will respond to upsets.
- Startups will go more smoothly and quickly, especially if control strategy design includes making estimates of controller tuning.
- Control strategy and process design will be compatible with business goals. This will result in achieving the desired yields, capacity, cycle times, and so forth.

ASPECTS OF PROCESS DESIGN THAT ADVERSELY AFFECT CONTROLLABILITY

There are a number of underlying causes of process variability that are introduced in process design. Improving controllability becomes a matter of recognizing and removing or minimizing the causes if they are present by the appropriate design changes. Some common causes are listed in Figure 10–1. The reader is also referred to a paper by Shinskey (1983) in which he analyzed several types of processes to show how uncontrollable operation can occur and what to do about it. Another good reference is the classic text by Buckley (1964).

Long Time Delays (Deadtime)

Time delay, or what control engineers refer to as deadtime, comes about when the controlled variable and the manipulated variable are separated in time. In other words, if a change is made in the manipulated variable, there is a time lapse before the controlled variable begins to change. The larger the delay, the more difficult it is to control by conventional control algorithms. For example, the integral mode in a PID controller continuously integrates the deviation between the setpoint and the process variable even if the variable is not changing. If the delay is long enough, the controller might keep integrating until it reaches its high or low limit (saturation). If you were controlling the process manually, you would probably wait until the process responded before making another adjustment so you could see the effect of the first adjustment. An ordinary PID has no such in-

> **Design Causes of Variability**
>
> Long Time Delays (Dead Time)
>
> Long Time Constants
>
> Inverse Response
>
> Nonlinearities
>
> Interaction
>
> Heat Integration
>
> Surge Capacity
>
> Large Turndown Requirements
>
> Equipment Susceptibility to Upsets

Figure 10–1 Variability that Can Be Linked to Process Design

telligence, but other means have been devised to make controllers more intelligent. Thus, if you recognized where a long time delay existed in the design, you could either try to minimize it by changing the design or decide to install a controller that could cope with the delay.

An obvious way to reduce deadtime is to locate the process variable measurement close to the manipulative variable. Relocating the measurement point may require a change in design to add a new nozzle or relocate an old one.

In the case where a lab measurement is involved in causing a long time delay, the engineer might decide that an on-line predictive model is worth developing; or perhaps a process variable can be found that is a good indicator of the property being measured in the lab.

Inverse Response

Inverse response is where the process initially responds in a direction that is opposite the steady-state change. Inverse response occurs in some distillation column base-level control loops where the level is controlled by manipulating vapor boilup. If the level increased, you would increase boilup to drive more material out of the base. However, with some types of trays, an increase in boilup momentarily dumps more liquid from the trays into the downcomer, which results in more fluid entering the base than is boiled off, thus increasing the level. This confuses the controller, which expects the base level to drop. This momentary phenomenon can cause a control loop to become unstable. Once this is recognized, the control engineer can either change the strategy or perhaps install a compensator for inverse response [Buckley et al. 1985].

Nonlinearities

Since conventional controllers are linear and tuned for a specific process response to a change in the manipulated variable, the degree of stability is compromised if the process is nonlinear and exhibits changing process gains, time constants, and deadtimes.

One type of nonlinearity is where a process variable has a different time response depending on the direction of the manipulated variable, particularly in the magnitude of the response. This occurs frequently in high-purity distillation columns in which a product composition is being controlled. If the bottom purity is being controlled by manipulating boilup, an increase in boilup may increase purity very slightly; but a decrease in boilup of the same magnitude may decrease purity by a tremendous amount in comparison. Recognizing that this can occur, the control engineer would recommend that a more linearly responding process variable be controlled instead. This could affect the location of the measurement, and therefore the nozzles on the equipment.

Waste treatment systems can also be nonlinear in the process gain depending on the amount of buffering. Knowing that these nonlinearities exist, the control engineer would suggest a specific design for the process vessels to improve control.

Interaction

Process interactions occur when the action of one controller affects another controlled variable or when the response of one variable affects the response of another. Again, using the example of a distillation column, controlling the product purities at two locations often leads to interactions that cause increased variability and even unstable operation when both controllers are in service. This may call for the design of interaction compensation or more advanced control (e.g., model predictive control).

Another example is where equipment is arranged and operated in parallel (e.g., dual heaters in a Dowtherm system). This design often calls for complex control loops involving multiple valves. Recognizing that a potential problem exists during the design stage, the design can be made to accommodate the multiple valves and extra piping.

Heat Integration

The use of process-to-process heat exchangers may require bypasses and special control strategies to make the system stable or facilitate startup. Another heat integration scheme that requires special design considerations is when the vapor from one distillation column provides the boilup for a second column. The correct control strategy is generally not obvious and may require simulation to find an appropriate one. A control engineer can recognize the critical nature of such designs and the special design requirements for both steady-state operation and startup. For example, it may be necessary to provide auxiliary condensers or reboilers to enable startup or to be able to operate over the full range of production rates.

Surge Capacity

The interactive nature of continuous processes often dictates that extra consideration be given to smoothing flows between vessels to minimize variations. The control engineer can recognize when these situations exist and recommend vessel sizes that allow the level controls to be tuned slowly, letting the level drift about the setpoint. That way, surges in flow are absorbed and not passed on to the next vessel. Typically, distillation column bases and condensate tanks are sized to achieve flow smoothing when several columns are operated in series [Buckley et al. 1985].

On a recent project, we were constrained in the additional holdup we could gain in a column base inventory and knew that the surge capacity was marginal. In this situation, the base level was controlled by manipulating boilup because the tail flow was an intermittent purge. The distillate flow from this column fed a second column, which had tight specifications on its products. As expected, we found at startup that the base level in the first column was not adequate to dampen the changes in vapor boilup, and the feed to the second column was more oscillatory than desirable. This strongly confirms how important flow smoothing is for minimizing variability under automatic control.

Long Time Constants

The time constant is, roughly speaking, the time it takes a process variable to reach 63% of its steady-state value along an exponential trajectory following a step change in an input variable (e.g., manipulative variable). This exponential time response is normally referred to as a *lag*. A long time constant in a control loop may be incompatible with tight control of variability, and design changes may be necessary to speed up the response. For example, long time constants can arise when trying to control a reactor temperature by manipulating cooling water to the reactor jacket. Thick reactor walls introduce long time constants. It may then be necessary to add internal coils or provide a recirculating cooling system to increase the heat transfer capability. The dynamics can be determined by dynamic analysis or simulation to show if this is a problem.

One example where time constants were important was in a gas phase reaction taking place in a tubular reactor. It was desirable to control the exit temperature because it was a good indicator of the degree of completion of the reaction and the concentration of exiting gases. We found out by computer modeling that the conventional method of measuring temperature by welding a thermocouple to the reactor wall was too slow, so we ended up installing the thermocouple directly in the gas at the exit of the reactor. Then we manipulated feed gas flow rate to control exit temperature. This turned out to be a very fast loop, and we were able to minimize variability in exit temperature.

Recycles also introduce time constants that may slow down the speed of response of process variables. In some cases, recycle lags may be minimized by changing the design or at least compensated for by the control strategy.

Large Turndown Requirements

Turndown is the ratio of maximum design capacity to minimum design capacity. This is not always dealt with adequately in process design in terms of what it means to equipment sizing and operating procedures. Quite often, if a plant is expected to run over a wide range of production rates (> 4:1), we may have to provide parallel valves, one large and one small, to get the required range.

Equipment that must operate over wide ranges also introduces the potential to operate equipment in regimes that result in poor control and even instability. For example, Buckley (1964) points out that in heat exchangers where a liquid is being boiled, it is possible to exceed the critical temperature difference between the hot and cold fluids. When this happens, the mode of heat transfer switches from nucleate boiling to film boiling, and the heat transfer rate *declines* as the temperature difference increases, which is exactly the opposite of normal operation. The result may be an unstable control loop because the controller action is set up for normal operation.

Another process problem caused by large turndown involves a condenser where the vapor temperature is greater than the boiling point of the coolant fluid. At low rates, the temperature of the coolant may be heated above its boiling point.

Finally, large turndown can cause process parameters like deadtime, time constants, and gains to change as the production rates change. This may require retuning the controllers to maintain the same level of stability. If tuning changes often, the engineer may recommend that an adaptive controller be applied that automatically retunes itself.

Equipment Susceptibility to Upsets

Certain kinds of process equipment are particularly susceptible to upsets. One example is an air-cooled condenser. It can be a real problem if it condenses overhead vapors from a distillation column and part of the condensate is returned as reflux to the column. The air-cooled condenser is susceptible to sudden ambient upsets like rain showers or sudden cold fronts. The rain dramatically increases the heat transfer coefficient and the condensate rate, causing wild pressure swings in the column. A cold front increases the temperature difference and therefore the condensate rate, and reduces the reflux temperature. This increases *internal* reflux by condensing more of the vapor in the column, thus upsetting the composition profile. One way to compensate for the colder reflux is to calculate and control internal reflux flow instead of the external flow. The calculation is simply a heat balance that requires knowledge of the vapor and condensate temperatures and the external reflux flow. The control engineer can spot when this may be a danger and recommend that nozzles be added to facilitate the additional measurements.

STRATEGY FOR REVIEWING DESIGN FOR CONTROLLABILITY

The control engineer should be part of the design team and have responsibility for reviewing the process design from the viewpoint of controllability implications and their effect on meeting the business goals. This task requires a good understanding of process dynamics, that is, how process equipment and configurations will respond over time to various upset conditions. If undesirable dynamic behavior potentially exists, the engineer should devise strategies for overcoming the problem and recommend them to the project team. Process understanding is key here; if it is lacking, consideration should be given to using modeling and simulations to gain it. The following steps can be taken to review the process and conceptualize the control strategies.

Material Balance Controls

The first consideration should be controlling the main liquid and vapor flows through the process. Buckley (1964) suggests starting at the finished product end and then, following the primary flows back to the raw material storages, design a continuous chain of material balance controls. Material balance controls will be made up of level and flow controls for liquids, and pressure and flow controls for gases. Particular attention should be paid to flow smoothing and the associated surge capacity requirements. Often, intermediate storages can be eliminated by increasing the holdup times in vessels such as the bases of distillation columns.

There are basically two ways material balance can be achieved: by setting the feed flows at the front end and *pushing* the flows forward, or by setting the product flows at the back end and *pulling* the flows forward. Buckley (1964) points out that the advantages of the latter method are that smaller inventories and less turndown capability are needed.

Next, the material balance controls for the secondary flows are reviewed. Normally, these will be manipulated in proportion to the primary flows either directly, by flow ratio controls, or indirectly, by quality controls. Special consideration may have to be given to automatic startups and shutdowns, transitions, and so forth.

Quality Controls

The manipulated variables not assigned to material balance strategies are available to control product quality. Going through the process, the control engineer should note every place where product quality is affected and the kinds of upsets that can occur. Then control strategies are designed to provide quality control at each key location. Note that we also include the control of process variables that ultimately affect quality, such as concentration, temperature, and so on. This is also the time to note any special kinds of sensors or analyzers that are needed. The strategies should try to avoid all the troublesome factors noted earlier (e.g., process deadtime).

Auxiliary Controls

Last, miscellaneous control needs are addressed. Examples are such things as maintaining process temperatures, staying within constraints, and controlling various utilities that have not already been covered in either the material balance controls or the quality controls.

FOLLOWUP WITH THE DESIGN TEAM

When the process design is complete and the controls have been conceptualized, it is time to review the recommendations with the process design team to take advantage of their experience and insights, and perhaps get additional information that was not initially available. If there are unanswered questions resulting from a lack of process understanding, this is a good time to get agreement on how that understanding will be gained, through additional resources or perhaps by computer simulation. The estimated changes in vessel sizes are also agreed upon.

Once the project team agrees on the design concepts, the control engineer does whatever modeling and simulation is required to finalize design, finishes the calculations for the vessel holdup volumes, and estimates control tuning parameters and other calibration factors. An explanation of the control strategy, including sketches and calculations, should be formally documented for inclusion in the process design manual. This information will be used in developing operating procedures and training programs.

The control engineer can add further value to the project by following the design through to completion to make sure the recommendations are properly implemented. The engineer will also be a valuable member of the commissioning team to aid in fine-tuning the controls at startup and getting the process operating smoothly.

EXPERT SYSTEMS FOR PROCESS CONTROL DESIGN

Computers have been used for many years to aid in process design. There are numerous steady-state programs for calculating the number of trays in distillation columns, flow and pressure drops in piping systems, the number of tubes in heat exchangers, and even complete flow sheets. It is difficult to imagine how we could design and build plants today without the aid of computers. We also make

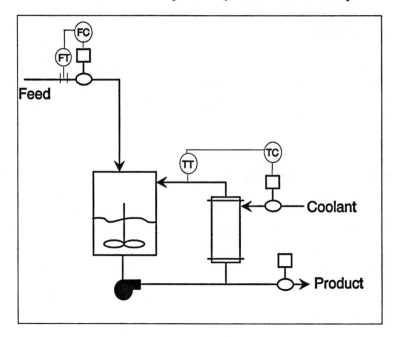

Reaction Rate Controls on a Continuous Reactor
with Recirculating Contents

*The reactor contents are recirculated through an external
heat exchanger to which a cooling medium is added. A
reactor temperature control adjusts the coolant valve. This
strategy is preferred over coolant jacket because of the
greater speed of response.*

Select other control variables to complete the strategy.

Figure 10–2   The Control Strategy for Exothermic CSTR Recommended by
the Expert System

a lot of use of dynamic simulations to predict the dynamic response of processes
to disturbances and check out control strategies and controller tuning.

We are now seeing more in the way of expert systems to aid in the design
of processes and equipment. There are a number of expert systems used to design
process control strategies. A number of references are cited for the design of dis-

tillation column control strategies alone [Shinskey 1986, Birky et al. 1988, Jensen et al. 1990]. These expert systems determine the appropriate pairing of controlled and manipulated variables to provide a controllable strategy. One of the major benefits of this technology is, of course, the leveraging of expertise, since so many companies have reduced technical forces to the bare minimum.

Because there are savings to be gained by the proper control of many key pieces of equipment, particularly in reaction systems, expert systems that aid in selecting appropriate control strategies for the total plant are also very useful and beneficial. One such expert system, developed by the author and his coworkers, will be briefly described.

The expert system is based on a commercial expert system shell and is currently implemented on a VAX computer that is a node on the company computer network. Although it could also be implemented in a PC, the VAX was chosen so that a wide group of engineers across the company could have free access to the program. The expert system has two purposes:

- To assist in the selection of control strategies for new or modernized facilities
- To provide a reference against which to compare existing strategies for troubleshooting purposes

A large selection of unit operations are included, from absorbers to waste water treatment systems. The expert system is structured so that the engineer selects the desired unit operation, and then, based on the responses to a series of questions, specifies the particular equipment configuration, desired controlled variables, and other details to define the particular application. The expert system presents a schematic diagram of the recommended control strategy and a narrative describing how the strategy works and why it was chosen.

Below is an example of selecting a reactor control strategy. The selections made by the engineer are indicated by the arrows (>).

```
TYPE OF REACTOR:
  > CONTINUOUS
  BATCH
CONTROLLED VARIABLE:
  > REACTION RATE
  PRESSURE
  FEED FLOWS
```

TYPE OF VESSEL:
> CSTR
TUBULAR OR FIXED BED
TYPE OF REACTION:
ENDOTHERMIC
> EXOTHERMIC
HEAT TRANSFER SYSTEM:
COIL/JACKET
> RECIRCULATING REACTOR CONTENTS
RECIRCULATING HEATING/COOLING MEDIUM

Once the engineer has specified the application details, the expert system recommends a control strategy, such as that in Figure 10–2. The engineer checks the design of the reactor to make sure it represents the correct application, then applies the control strategy to the process drawings and so on. Conceivably, the expert system could be integrated into the computer-aided design system so that control strategy design could be automated along with P&I drawings.

REFERENCES

BIRKY, G. J., MCAVOY, T. J., and TYREUS, B. D. *Expert System for Design of Distillation Controls. ISA Proceedings,* Houston, October 1988.

BUCKLEY, P. S. *Techniques of Process Control.* New York: John Wiley, 1964.

BUCKLEY, P. S., LUYBEN, W. L., and SHUNTA, J. P. *Design of Distillation Column Control Systems.* Research Triangle Park, NC: Instrument Society of America, 1985.

BUCKLEY, P. S. *Process Control Strategy and Profitability.* Research Triangle Park, NC: Instrument Society of America, 1992.

JENSEN, B., HARPER, T., and LIKINS, M. *An Expert System in Fractionator Control System Design.* Paper presented at the A.I.Ch.E. Spring Meeting, Orlando, FL, 1990.

SHEFFIELD, R. E. Integrate Process and Control System Design. *Chemical Engineering Progress,* pp. 30–35, October 1992.

SHINSKEY, F. G. Uncontrollable Processes and What to Do About Them. *Hydrocarbon Processing,* pp. 179–182, November 1983.

SHINSKEY, F. G. An Expert System for the Design of Distillation Controls. *Proceedings of the Third International Conference on Chemical Process Control,* Asilomar, CA, November 1987.

INDEX